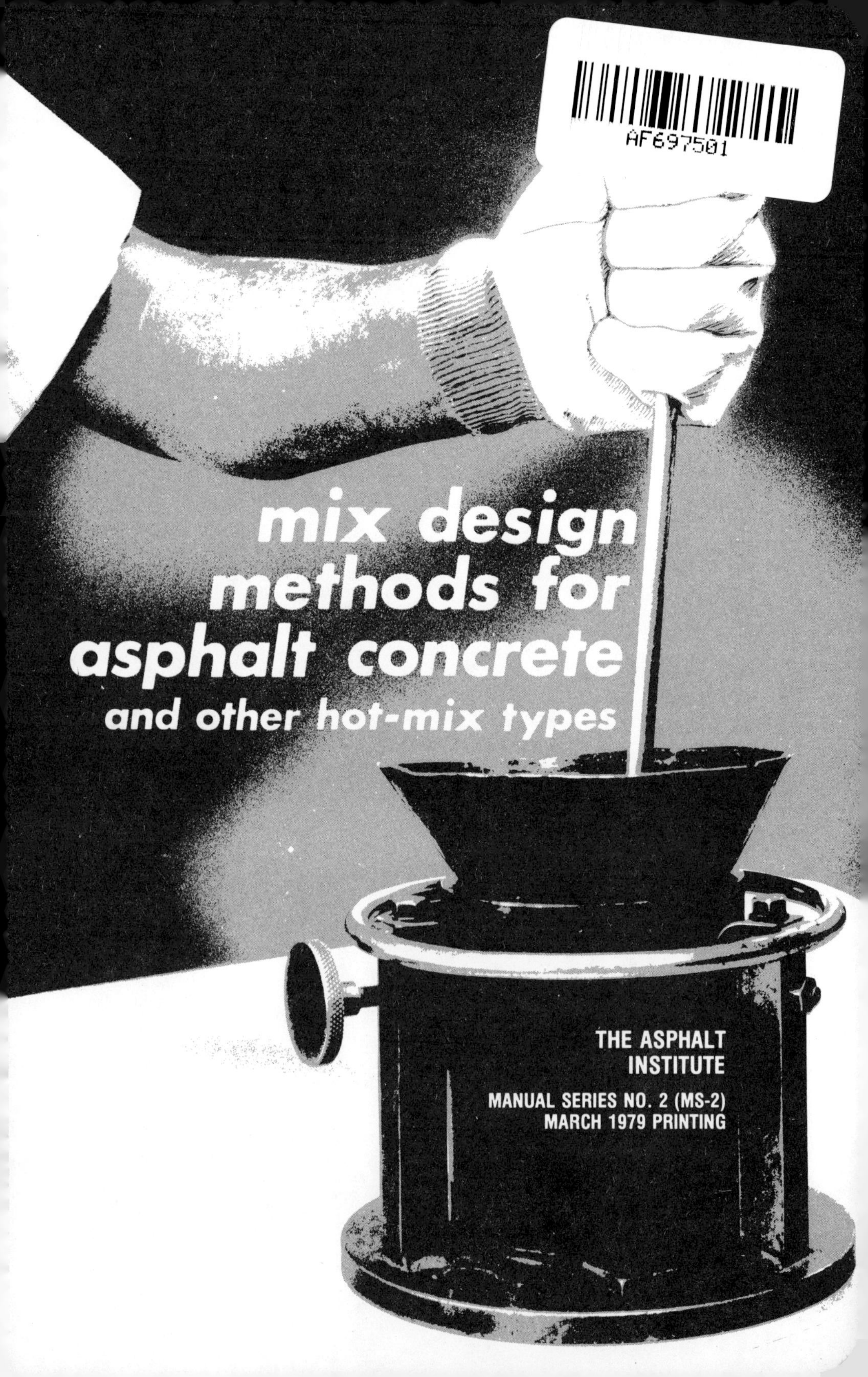

mix design methods for asphalt concrete

and other hot-mix types

THE ASPHALT INSTITUTE

MANUAL SERIES NO. 2 (MS-2)
MARCH 1979 PRINTING

FOREWORD

This edition is the result of revisions reflecting advances in mix design methods for asphalt mixes.

It is intended as a practical guide to asphalt mix design for the use of engineers and all concerned with the technicalities of constructing highways and other structures with asphalt.

Its use as a textbook for students is eminently practical, and is recommended.

It is by no means in its final form, since continuing research undoubtedly will provide advances in mix design that will be included in future revised editions.

The Asphalt Institute welcomes comments on this publication.

THE ASPHALT INSTITUTE
Asphalt Institute Building
College Park, Maryland 20740
March 1979

CONTENTS

ILLUSTRATIONS

ILLUSTRATIONS (Continued)

TABLES

Chapter I

INTRODUCTION

1.01 PURPOSE AND SCOPE–With the aid of modern scientific research and construction methods, the use of asphalt in highway construction is expanding rapidly.

Hot-mixed hot-laid asphalt paving mixtures, long considered by many highway engineers as the premier paving products available anywhere at any cost, are becoming America's most popular paving materials.

A number of engineers and construction organizations have expressed a need for sound engineering data for the instruction of their design and construction personnel in the use of current hot-mix design methods. This manual has been written with this need in mind. It contains the latest information for the design of hot-mix paving mixtures to meet the demands of modern traffic conditions and to ensure top performance of asphalt pavement.

Chapter II of this manual reviews the application of design testing to general construction practice. Testing references and detailed procedures are outlined for the routine analysis of materials and paving mixtures. A number of typical examples of routine computations and calculations related to mix design are included in the Appendix. The principal feature of this manual, however, is the detailed outline of test methods and procedures for two methods of asphalt paving mix design:*

The Marshall Method
The Hveem Method

In each instance, the method of design and the test criteria are presented free from specification requirements for materials and construction.

The Marshall and Hveem methods of mix design are widely used for the construction design of hot-mix paving. The selection and use of either of these mix design methods is principally a matter of engineering preference, since each method has certain unique features and

*The Hubbard-Field Method, included in earlier editions of this manual, has been deleted because of its limited use. The test method, however, is published by the American Society for Testing and Materials as Standard Method of Test D 1138 and by the American Association of State Highway Officials as Standard Method of Test T 169.

apparent advantages. Both methods are currently being used with satisfactory results.

1.02 HOT-MIX DEFINED—Hot-mixed asphalt paving consists of a combination of aggregates uniformly mixed and coated with asphalt cement. To dry the aggregates and obtain sufficient fluidity of the asphalt cement for proper mixing and workability, both must be heated prior to mixing—giving origin to the term "hot-mix."

The aggregates and asphalt are combined in an asphalt mixing plant in which they are heated, proportioned, and mixed to produce the desired paving mixture. After the plant mixing is complete, the hot-mix is transported to the paving site and spread with a paving machine in a loosely compacted layer to a uniform, smooth surface. While the paving mixture is still hot, it is further compacted by heavy self-propelled rollers to produce a smooth, well-consolidated course.

1.03 CLASSIFICATION OF HOT-MIX ASPHALT PAVING—Asphalt paving mixes may be designed and produced from a wide range of aggregate blends, each suited to specific uses. The aggregate composition may vary from coarse to fine particles. Many different compositions are specified throughout the world—the ones in any given locality generally being those that have proven adequate through long usage and, in most cases, these should be used. For a general classification of mix compositions, however, The Asphalt Institute recommends consideration of mix designations and nominal maximum size of aggregate (2A) 1-1/2 in., (3A) 1 in., (4A) 3/4 in., (5A) 1/2 in., (6A) 3/8 in., (7A) No. 4, and (8A) No. 16 as specified in ASTM* Standard Specification D 3515 for Hot-Mixed, Hot-Laid Bituminous Paving Mixtures. The grading and asphalt content limits of these mixes agree, in general, with overall practice but may vary from the practice of a given local area.

---

*American Society for Testing and Materials

Chapter II

MIX DESIGN PRACTICE

2.01 GENERAL–Asphalt paving mix design demands attention to the details of design test procedures. Mainly, this means following written instructions. But it also means having proper training in laboratory technique and the relation of mix design testing to the specification requirements.

While mix design often is treated as a subject in itself, it cannot be divorced from the other related items of the specifications. It is the purpose of this chapter, therefore, to cite the general objectives of mix design and present a guide for applying mix design procedures to paving construction specifications.

2.02 OBJECTIVES OF ASPHALT PAVING MIX DESIGN–The design of asphalt paving mixes, as with other engineering materials designs, is largely a matter of selecting and proportioning materials to obtain the desired properties in the finished construction. *The overall objective for the design of asphalt paving mixes is* to determine an *economical blend and gradation of aggregates* (within the limits of the project specifications) *and asphalt* that yields a mix having:

(1) *Sufficient asphalt* to ensure a durable pavement.

(2) *Sufficient mix stability* to satisfy the demands of traffic without distortion or displacement.

(3) *Sufficient voids* in the total compacted mix to allow for a slight amount of additional compaction under traffic loading without flushing, bleeding, and loss of stability, yet low enough to keep out harmful air and moisture.

(4) *Sufficient workability* to permit efficient placement of the mix without segregation.

2.03 EVALUATION AND ADJUSTMENT OF MIX DESIGNS–Often, in the process of developing a specific mix design, it is necessary to make several trial mixes to find one that meets all the criteria of the design method used. Each trial mix design, therefore, serves as a guide for evaluating and adjusting the trials that follow. For preliminary or exploratory mix designs it is advisable to start with an aggregate of a gradation that approaches the median of the specification limits. Initial trial mixes for establishing the job-mix formula, however, must have an aggregate gradation within the specification limits that the central mixing plant is producing or is capable of producing. *(See also* Article 2.05 Coordination of Mix Design Testing that follows in this chapter.)

Where the initial trial mixes fail to meet the design criteria it will be necessary to modify or, in some cases, redesign the mix. Adjustments in the grading of the original aggregate blend will be required to correct the deficiency.

For many engineering materials, the strength of the material frequently is thought of as denoting quality; however, this is not necessarily the case for hot-mixed asphalt paving. Extremely high stability often is obtained at the expense of lowered durability, and vice versa. Therefore, in evaluating and adjusting mix designs always keep in mind that the aggregate gradation and asphalt content in the final mix design must strike a favorable balance between the stability and durability requirements for the use intended. Moreover, the mix must be produced as a practical and economical construction operation.

Grading curves are helpful in making necessary adjustments in mix designs. For example, curves determined from the Fuller equation* represent maximum density and minimum voids in mineral aggregate (VMA) conditions. Paving mixtures with such curves tend to be workable and readily compacted. However, their void contents may be too low. Usually, deviations from these curves will result in lower densities and higher VMA. The extent of change in density and VMA depends on the amount of adjustment in fine or coarse aggregate. Figure II-1 illustrates a series of Fuller maximum density curves plotted on a conventional semilog grading chart.

Figure II-2 illustrates maximum density curves determined from the maximum density equation raised to the 0.45 power† and plotted on the Federal Highway Administration grading chart (based on a scale raising sieve openings to the 0.45 power), which many designers find convenient to use for adjusting aggregate gradings. The curves on this chart, however, need not be determined from the maximum density equation. They may be obtained by drawing a straight line from the origin at the lower left of the chart to the desired nominal maximum particle size‡ at the top. Gradings that closely approach this straight line usually must be adjusted away from it within acceptable limits to

---

*$p = 100(d/D)^{.5}$ in which p = total percentage passing given sieve
d = size of sieve opening
D = largest size (sieve opening) in gradation

†$p = 100(d/D)^{.45}$

‡For processed aggregate, the nominal maximum particle size is the largest sieve size listed in the applicable specification upon which any material is permitted to be retained.

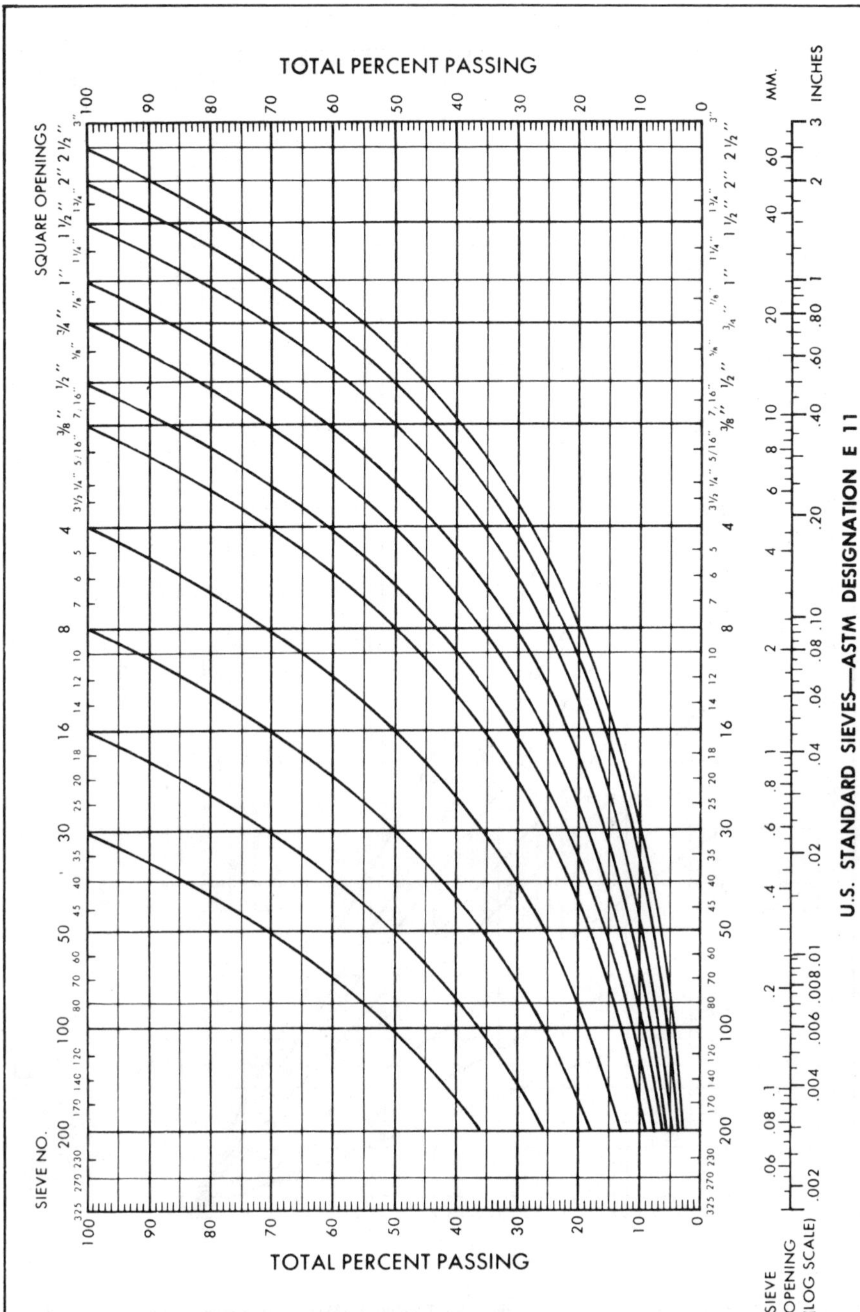

Figure II-1 – Fuller maximum density curves on standard semi-log grading chart

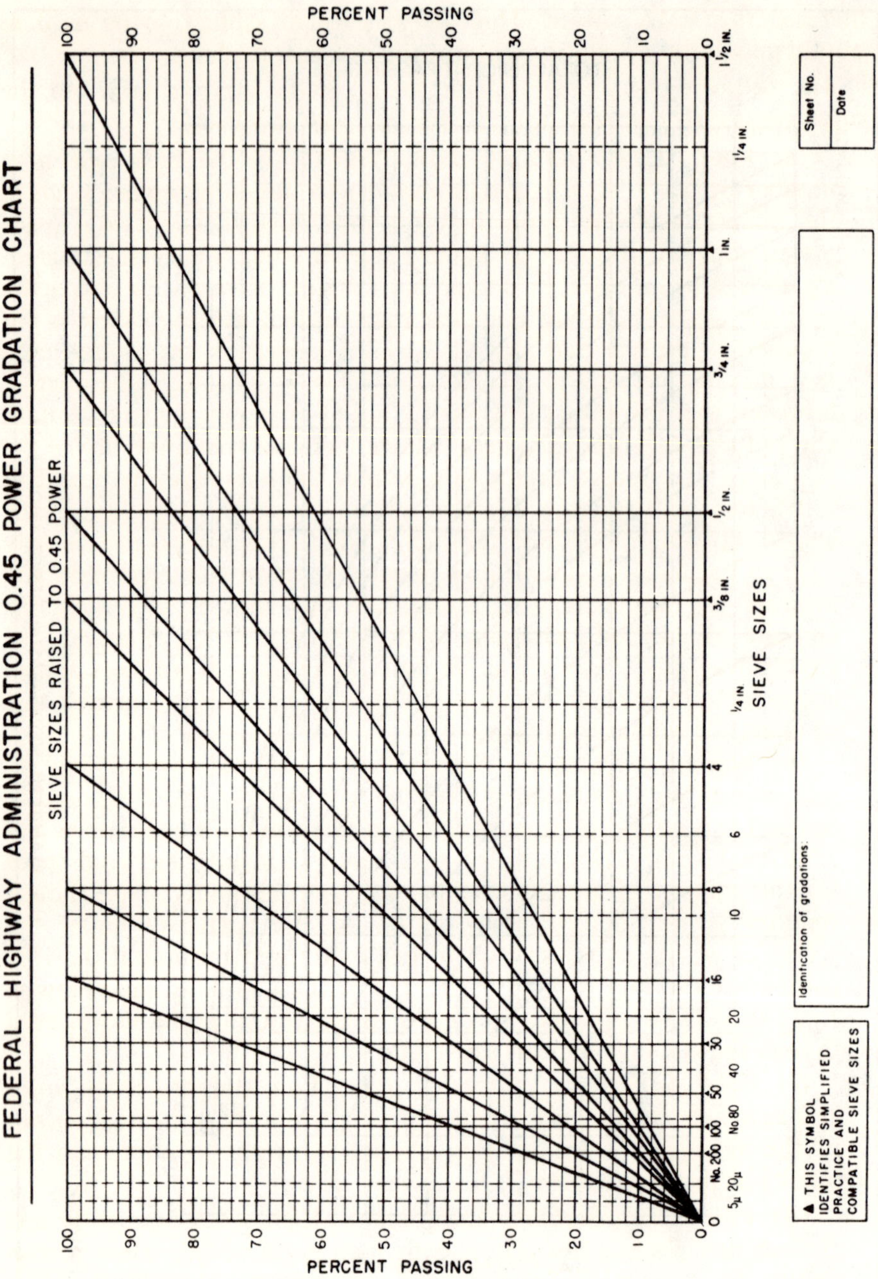

Figure II-2–Maximum density curves on Federal Highway Administration 0.45 power gradation chart

increase the VMA values. This allows enough asphalt to be used to obtain maximum durability without the mixture flushing.

The following is a general guide for adjusting the trial mix, but the suggestions outlined may not necessarily apply in all cases:

(a) *Voids Low, Stability Low*—Voids may be increased in a number of ways. As a general approach to obtaining higher voids in the mineral aggregate (and therefore providing sufficient void space for an adequate amount of asphalt and air voids) the aggregate grading should be adjusted by adding more coarse or more fine aggregate.

If the asphalt content is higher than normal and the excess is not required to replace that absorbed by the aggregate, the asphalt content may be lowered to increase the voids. It must be remembered, however, that lowering the asphalt content increases the void content and reduces the film thickness, which decreases the durability of the pavement. Too great a reduction in film thickness also may lead to brittleness, accelerated oxidation, and increased permeability. If the above adjustments do not produce a stable mix, the aggregate may have to be changed. It usually is possible to improve the stability and increase the aggregate void content of the mix by increasing the amount of crushed materials. With some aggregates, however, the freshly fractured faces are as smooth as the water worn faces and an appreciable increase in stability is not possible. This is generally true of quartz or similar rock types.

(b) *Voids Low, Stability Satisfactory*—Low void content may result in instability or flushing after the pavement has been exposed to traffic for a period of time because of reorientation of particles and additional compaction. It also may result in insufficient void space for the amount of asphalt required for high durability, even though stability is satisfactory. Degradation of the aggregate under the action of traffic may also lead to instability and flushing if the void content of the mix is not sufficient. For these reasons, mixes low in voids should be adjusted by one of the methods given above, even though the stability appears satisfactory.

(c) *Voids Satisfactory, Stability Low*—Low stability when voids and aggregate grading are satisfactory may indicate some deficiencies in the aggregate. Consideration should be given to improving the quality as discussed above.

(d) *Voids High, Stability Satisfactory*—High voids frequently are, though not always, associated with high permeability. High permeability, by permitting circulation of air and water through the pavement, may lead to premature hardening of the asphalt. Even though stabilities are satisfactory, adjustments should be made to reduce the voids. This usually may be accomplished by increasing the mineral dust content of the mix. In some cases, however, it may be necessary to select or combine aggregates to more closely approximate the gradation of a maximum density grading curve.

(e) *Voids High, Stability Low*—Two steps may be necessary when the voids are high and the stability is low. First the voids are adjusted by the methods discussed above. If this adjustment does not also improve the stability, the second step should be a consideration of aggregate quality as discussed above.

2.04 DESIGN METHOD AND REQUIREMENTS—One should recognize at the outset that the mix design method and design requirements form an *essential* part of the construction specifications for all heavy-duty asphalt paving.

The construction agency or authority in responsible charge of the paving construction usually establishes, as a matter of policy, the mix design method and design requirements. Once these items are established, it then becomes the duty and responsibility of the engineer and inspector to develop the mix design within the framework of the other specification requirements.

The Marshall and Hveem methods of mix design presented in this manual have been widely used with satisfactory results. For each of these methods, criteria have been developed by correlating the results of laboratory tests on the compacted paving mixes with the performance of the paving mixes under service conditions. In each instance, however, the correlation was made within certain limits; these limits are clearly set forth for each method.

> IMPORTANT: For the above reasons, the design criteria for each mix design method are applicable only to the prescribed test procedure within the limits of the original correlation. *Hasty or haphazard modification of these design methods, test procedures or design criteria is never justified.* In those cases where it can be clearly shown that a modification or extension of the design method is needed, *all proposed changes should be fully supported with additional correlation data covering the new limits or conditions of design.*

2.05 COORDINATION OF MIX DESIGN TESTING—Mix design testing for asphalt paving construction must be thoroughly coordinated with the other features of materials and construction inspection. Normally, mix design testing will have four important applications to the overall project construction. These are: preliminary design, source acceptance, job-mix control, and routine construction control. The following outlines the purpose and objective for each application. This outline is intended to show *only* the relation of mix design testing to the overall program of inspection control. For the actual details of field inspection refer to The Asphalt Institute's *Asphalt Plant Manual* (MS-3) and *Asphalt Paving Manual* (MS-8).

(a) *Preliminary Design Testing*—The principal purpose of preliminary design testing is to determine that the local prospective sources of aggregate are of satisfactory quality and will produce a paving mix satisfying both the gradation and mix design requirements proposed for the specifications. When scheduling preliminary mix designs, it should first be determined that both asphalt and aggregate materials meet the proposed requirements of the specifications. Initial trial mixes should be made with an aggregate gradation approaching the median of the specification limits. Where there are several possible sources of aggregate to be considered, it will often be necessary to make a number of trial mix designs to determine the most economical combination of aggregates that will fulfill all of the specification requirements. The results of the preliminary mix designs serve as a basis for making a preliminary estimate of costs; they further assure the owner that the design requirements can be obtained within the framework of the specification. The latter is particularly important where there are no service records available on the prospective aggregate sources.

(b) *Source-Acceptance Testing*—Source-acceptance testing often is performed after the award of contract and after the contractor has indicated the proposed sources of all aggregate materials and asphalt. While this testing is primarily a matter of materials control and inspection, it also frequently involves mix design. The main objective is to determine the most economical blend of aggregates that will satisfy both the gradation and mix design requirements of the specifications. Source-acceptance tests serve to assure the owner and contractor alike that the proposed sources of aggregate and asphalt will yield a paving mixture that satisfies all specification requirements and, further, will permit the contractor to proceed with the stockpiling of the required materials at the job site.

(c) *Job-Mix Control Testing*—Job-mix control testing is performed at the start of plant production and in conjunction with the calibration of the mixing plant for the job-mix formula. This is one of the key points of design control for the paving construction since the job-mix formula establishes the actual gradation and asphalt content to be obtained in the mixing plant. For this reason mix design testing must be thoroughly coordinated with overall inspection control. Initial trial mixes for job-mix control testing are usually made on aggregate samples lifted from the plant. The aggregate gradations must be within the specification limits and the plant must be capable of producing them as a practical operation. Emphasis must be placed on accurate and complete job-mix control testing. Such testing serves to determine whether the paving mixture produced under the job-mix formula meets all the requirements of the specifications. It further serves as a basis for all the routine construction control testing that follows.

(d) *Routine Construction Control Testing*–This testing is performed as a routine and periodic item of inspection during the paving construction. Representative samples of the hot paving mixture are lifted at the mixing plant and check-tested for design properties. (A random sampling method should be used if the hot-mix plant is operating under statistical quality control specifications.) The results of these check tests are compared with the job-mix control tests and the overall specification requirements. In those instances where irregularities occur and the limits of the job-mix formula are exceeded, appropriate corrections will be required at the plant. Occasionally, where the situation warrants, it may be necessary to re-evaluate and redesign the paving mixture. Until the plant is producing the desired mix consistently, a large number of tests may be necessary. When the job is running smoothly the number of tests may be reduced to the recommended minimum–two check tests per day. A control check test should be made, however, whenever there is a hint that the mix is outside the job tolerance. Although this type of mix design testing is routine in nature, it requires close attention. Its importance is emphasized by the fact that the results of the routine control tests serve as a basis for the final acceptance of the paving construction by the owner.

2.06 AGGREGATE GRADATIONS AND FRACTIONS–For the purpose of specifications and test reporting it is almost universal practice to specify the gradation of aggregates on the basis of the total aggregate gradation, i.e., total percent by weight passing the designated sieve sizes (*see* example below). The individual fractions of the total aggregate gradation, however, are designated as follows:

Coarse Aggregate (retained No. 8 sieve)
Fine Aggregate (passing No. 8 sieve)
Mineral Dust (passing No. 200 sieve).

It is also important to note that the aggregate gradations as well as the individual fractions are specified independently of the total mix; i.e., the total aggregate equals 100 percent.

Aggregate materials often are identified in broader terms as: rock, sand, and filler. These terms usually are applied to the stockpiled materials supplied to the job site. The following definitions appear to have the greatest usage:

Rock–Material that is predominantly coarse aggregate (retained No. 8).

Sand–Material that is predominantly fine aggregate (passing No. 8).

Filler–Material that is predominantly mineral dust (passing No. 200).

EXAMPLE

The following example is presented to illustrate application of the above items. On a certain paving job it is determined that the most economical sources of aggregate are crushed limestone for the coarse sizes, and river sand supplemented with limestone dust for the fine sizes. The gradation analysis on each of these three materials shows that they should be blended,

46 percent Rock (crushed limestone)
48 percent Sand (river sand)
6 percent Filler (limestone dust)
100 percent

to produce the required gradation in the total aggregate. (*See* Appendix for typical gradation and blending calculations.) The total gradation produced by the above blend proportions is as follows:

Sieve Sizes	Total Passing Percent by Weight	
	Specs.	Blend
3/4" (19.0 mm)	100	100
1/2" (12.7 mm)	90 – 100	95
#4 (4.76 mm)	45 – 70	62
#8 (2.38 mm)	25 – 55	40
#50 (0.297mm)	5 – 20	11
#200 (0.074mm)	2 – 9	6

2.07 PROPORTIONING AGGREGATE AND ASPHALT–In the proportioning of aggregate and asphalt it is important to note that the asphalt content may be expressed either as a percentage by weight of total mix or as a percentage by weight of dry aggregate. Each method of specifying asphalt has certain advantages and either method is acceptable provided it is clearly understood which method is being used. The following example illustrates the two methods:

EXAMPLE

Assume that a given mix contains aggregates and asphalt in the proportion of 100 lb (45.4 kg) of aggregate to 6 lb (2.72 kg) of asphalt. The asphalt content of such a mix could be expressed as (6 ÷ 100) 100 = 6 percent asphalt *by weight of dry aggregate*. This asphalt content may also be expressed [6 ÷ (100 + 6)] 100 = 5.66 percent asphalt *by weight of total mix.*

It is therefore important that the basis on which the asphalt content is expressed be clearly understood.

2.08 TESTING METHODS—The methods of testing the constituent materials for specification requirements will normally be set forth in the project specifications. In those cases where the testing methods are not defined, the following schedule of test methods is recommended. Test methods shall be the latest revision of methods adopted by the American Association of State Highway Officials or the American Society for Testing and Materials as follows:

(a) Asphalt Cement	AASHO Designation	ASTM Designation
Penetration	T 49	D 5
Viscosity		
Absolute	T 202	D 2171
Kinematic	T 201	D 2170
Flash Point	T 48	D 92
Thin Film Oven Test	T 179	D 1754
Rolling Thin Film Oven Test	T 240	D 2872
Ductility	T 51	D 113
Solubility	T 44	D 2042
Specific Gravity	T 228	D 70

(b) Mineral Aggregates	AASHO Designation	ASTM Designation
Percent Wear	T 96	C 131
Unit Weight	T 19	C 29
Sieve Analysis (Aggregates)	T 27	C 136
Sieve Analysis (Filler)	T 37	D 546
Specific Gravity (CA)	T 85	C 127
Specific Gravity (FA)	T 84	C 128
Specific Gravity (Filler)	T 100 or T 133	D 854 or C 188
(c) Hot-Mix Asphalt Paving		
Bitumen Content	T 164	D 2172
Recovery of Asphalt	T 170	D 1856
Density and Voids Analysis	*See* Chapter V	
Maximum Specific Gravity of Paving Mixtures	T 209	D 2041
Bulk Specific Gravity	T 166	D 1188 or D 2726

2.09 SCHEDULE OF SAMPLES AND TESTS—Prior to mix design testing, be sure that *representative* samples of aggregates and asphalt are in ample supply to accomplish the required number of tests. The following schedule of quantities is suggested:

Asphalt Cement	1 gal (4 litres)
Coarse Aggregate (or Rock)	50 lb (23 kg)
Fine Aggregate (or Sand)	50 lb (23 kg)
Filler (when required)	20 lb (9 kg)

Additional quantities of aggregate may be required if the above quantities will result in appreciable waste in combining materials for the design gradation or if several combinations of coarse and fine aggregates are to be investigated.

Each sample of material should be completely identified by source location, project location, and project number or job number. Each sample of asphalt cement should be in clean, small metal containers with tight lids or covers to avoid the necessity of reheating the entire supply each time a mix is tested. Each aggregate sample should be in a tightly woven cloth sack, securely wired or tied.

In advance of the mix design testing, a list or schedule of the tests to be performed should be made. Tests should be listed in their proper and logical sequence. It is important for all other specification tests to be completed before the mix design tests are started. In this way the sources of substandard aggregates are eliminated from the design studies. The Appendix, Gradation Analysis of Aggregates, suggests a schedule of aggregate analysis and testing.

2.10 PREPARATION OF TEST MIXES–Detailed procedures used in the preparation of test mixes for each method of mix design are outlined later in this manual. In general, however, the procedures illustrated in Figures II-3 through II-6 are used.

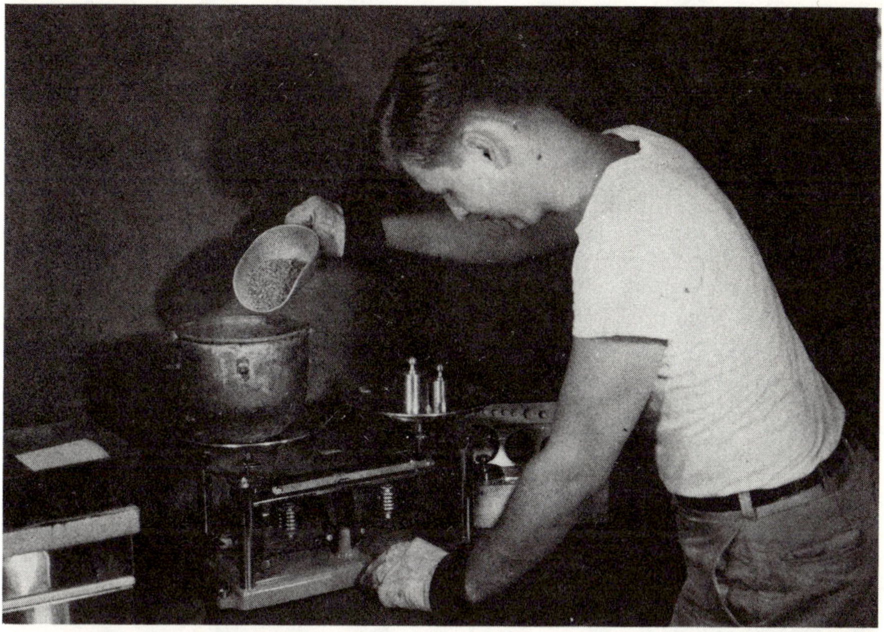

Figure II-3—Weighing aggregates for batch mixes

Figure II-4—Heating aggregate batches in oven

Figure II-5—Adding asphalt from a thermostatically-controlled asphalt heating kettle

Figure II-6—Mechanical mixer for batch mixing of asphalt and aggregate

Chapter III

MARSHALL METHOD OF MIX DESIGN

A. *General*

3.01 DEVELOPMENT AND APPLICATION–The concepts of the Marshall method of designing paving mixtures were formulated by Bruce Marshall, formerly Bituminous Engineer with the Mississippi State Highway Department. The U.S. Corps of Engineers, through extensive research and correlation studies, improved and added certain features to Marshall's test procedure, and ultimately developed mix design criteria. The Marshall test procedures have been standardized by the American Society for Testing and Materials. Procedures are given by ASTM Designation D 1559, *Resistance to Plastic Flow of Bituminous Mixtures Using Marshall Apparatus*. Testing procedures presented here are basically the same as those of the ASTM Method.

The Marshall method as presented here is applicable only to hot-mix asphalt paving mixtures using penetration or viscosity graded asphalt cements and containing aggregates with maximum sizes of 1 inch (25 mm) or less. The method may be used for both laboratory design and field control of asphalt hot-mix paving.

3.02 OUTLINE OF METHOD–The procedure for the Marshall method starts with the preparation of test specimens. Preliminary to this operation it is, of course, required:

(a) That the materials proposed for use meet the requirements of the project specifications.

(b) That aggregate blend combinations meet the gradation requirements of the project specifications.

(c) That, for use in density and voids analyses, the bulk specific gravity of all aggregates used in the blend, and the specific gravity of the asphalt cement, are determined.

These requirements are matters of routine testing, specifications, and laboratory technique that must be considered but that are not unique to any particular mix design method. The engineer should refer to the Appendix, Gradation Analysis of Aggregates, for the schedule of preparation and analysis of aggregates.

The Marshall method uses standard test specimens of 2-1/2 inches (64 mm) height x 4 inches (102 mm) diameter. These are prepared using a specified procedure for heating, mixing, and compacting the asphalt-aggregate mixtures. The two principal features of the Marshall method of mix design are a density-voids analysis and a stability-flow test of the compacted test specimens.

The stability of the test specimen is the maximum load resistance in pounds (N) that the standard test specimen will develop at 140°F (60°C) when tested as outlined hereinafter. The flow value is the total movement or strain, in units of 1/100 inch (0.25 mm) occurring in the specimen between no load and maximum load during the stability test.

B. *Preparation of Test Specimens*

3.03 GENERAL–In determining the optimum asphalt content for a particular blend or gradation of aggregates by the Marshall method, a series of test specimens is prepared for a range of different asphalt contents so that the test data curves show a well-defined "optimum" value. Tests should be scheduled on the basis of 1/2 percent increments of asphalt content, with at least two asphalt contents above "optimum" and at least two below "optimum."

To provide adequate data, *triplicate* test specimens are the minimum prepared for each asphalt content used. Thus, a hot-mix design study using six different asphalt contents will normally require at least eighteen test specimens. Each test specimen will usually require approximately 1.2 kg of aggregate. Therefore, the *minimum* aggregate requirements for one series of test specimens of a given blend and gradation will be approximately 23 kg or approximately 50 lb. One gallon (4 litres approximately) of asphalt cement will be adequate.

3.04 EQUIPMENT–The equipment required for the preparation of test specimens is as follows:

(a) *Pans*, metal, flat bottom, for heating aggregates.

(b) *Pans*, metal, round, approximately 4-quart (4-litre) capacity, for mixing asphalt and aggregate.

(c) *Oven and Hot Plate*, electric, for heating aggregates, asphalt, and equipment as required.

(d) *Scoop*, for batching aggregates.

(e) *Containers*, gill-type tins, beakers, pouring pots, or sauce pans, for heating asphalt.

(f) *Thermometers,* armored, glass, or dial-type with metal stem, +50°F (10°C) to +450°F (232°C), for determining temperature of aggregates, asphalt and asphalt mixtures.

(g) *Balance*, 5-kg capacity, sensitive to 1 gm for weighing aggregates and asphalt. *Balance*, 2-kg capacity, sensitive to 0.1 gm for weighing compacted specimens.

(h) *Mixing Spoon*, large, or *Trowel*, small.

(i) *Spatula*, large.

(j) *Mechanical Mixer* (optional), commercial bread dough mixer 4-quart (4-litre) capacity or larger, equipped with two metal mixing bowls and two wire stirrers.

(k) *Boiling Water Bath*, consisting of hot plate and bucket for water, for heating compaction hammer and mold.

(l) *Compaction Pedestal*, consisting of an 8 x 8 x 18-inch (200 x 200 x 460 mm) wooden post capped with a 12 x 12 x 1-inch (305 x 305 x 25 mm) steel plate. The wooden post should be oak, yellow pine or other wood having a dry weight of 42 to 48 lb/ft^3 (673 to 769 kg/m^3). The wooden post should be secured by four angle brackets to a solid concrete slab. The steel cap should be firmly fastened to the post. The pedestal should be installed so that the post is plumb, the cap level, and the entire assembly is free from movement during compaction.

*(m) *Compaction Mold,* consisting of a base plate, forming mold, and collar extension. The forming mold has an inside diameter of 4 inches (101.6 mm) and a height of approximately 3 inches (76 mm); the base plate and collar extension are designed to be interchangeable with either end of the forming mold.

*(n) *Compaction Hammer*, consisting of a flat circular tamping face 3-7/8-inch (98.4 mm) diameter and equipped with a 10-pound (6.5 kg) weight constructed to obtain a specified 18-inch (457 mm) height of drop.

*(o) *Mold Holder*, consisting of spring tension device designed to hold compaction mold in place on compaction pedestal.

(p) *Extrusion Jack* or Arbor Press, for extruding compacted specimens from mold.

(q) *Gloves*, welders, for handling hot equipment. *Gloves*, rubber, for removing specimens from water bath.

(r) *Marking Crayons*, for identifying test specimens.

(Note: See additional equipment requirements in Article 3.07.)

3.05 PREPARATION OF TEST SPECIMENS–

(a) *Number of Specimens*–Prepare at least three, and preferably five, specimens for each combination of aggregates and asphalt content.

(b) *Preparation of Aggregates*–Dry aggregates to constant weight at 221°F (105°C) to 230°F (110°C) and separate the aggregates by dry-sieving into the desired size fraction. The following size fractions

*Marshall test apparatus should conform to requirements of ASTM test method D 1559.

are recommended as generally adequate: 1 to 3/4 inch (25.4 to 19.0 mm), 3/4 to 3/8 inch (19.0 to 9.52 mm), 3/8 to No. 4 (9.52 to 4.75 mm), No. 4 to No. 8 (4.75 to 2.36 mm) Passing No. 8 (2.36 mm).

(c) *Determination of Mixing and Compaction Temperature*—The temperature to which the asphalt must be heated to produce viscosities of 170 ± 20 centistokes kinematic and 280 ± 30 centistokes kinematic shall be established as the mixing temperature and compaction temperature, respectively.

(d) *Preparation of Mold and Hammer*—Thoroughly clean the specimen mold assembly and the face of the compaction hammer and heat them in a boiling water bath or on the hot plate to a temperature between 200°F (93°C) and 300°F (149°C). Place a piece of filter paper toweling cut to size in the bottom of the mold before the mixture is placed in the mold.

(e) *Preparation of Mixtures*—Weigh into separate pans for each test specimen the amount of each size fraction required to produce a batch that will result in a compacted specimen 2.5 ± 0.05 inches (63.5 ± 1.3 mm) in height. This will normally be about 1.2 kg. (*See* Appendix, Gradation Analysis of Aggregates, for suggested method of calculating batch weights.) It is generally desirable to prepare a trial specimen prior to preparing the aggregate batches. If the trial specimen height falls outside the limits, the amount of aggregate used for the specimen may be adjusted as follows:

Adjusted weight of aggregate =

$$\frac{2.5 \text{ (weight of aggregate used)}}{\text{Specimen height (in.) obtained}}$$ or, for International System of Units (SI),

Adjusted weight of aggregate =

$$\frac{63.5 \text{ (weight of aggregate used)}}{\text{Specimen height (mm) obtained}}$$

Place the pans in the oven or on the hot plate and heat to a temperature **approximately 50°F (28°C) above the mixing temperature** specified in (c). If a hot plate is used, provision should be made for dead space, baffle plate, or a sand bath beneath the pans and the hot plate to prevent local over-heating. Charge the mixing bowl with heated aggregates and dry mix thoroughly. Form a crater in the dry blended aggregate and weigh the required amount of asphalt cement into the mixture in accordance with the accumulative batch weights. At this point the temperature of the aggregate and asphalt must be within the limits of the mixing temperature established in paragraph (c). Asphalt should not be held at mixing temperatures for more than one hour before using. Mix the aggregate and asphalt cement, preferably with a mechanical mixer or by hand with a trowel, as quickly and thoroughly as possible to yield a mixture having a uniform distribution of asphalt throughout.

(f) *Compaction of Specimens*—Place the entire batch in the mold, spade the mixture vigorously with a heated spatula or trowel 15 times around the perimeter and ten times over the interior. Remove the collar and smooth the surface to a slightly rounded shape. The temperature of the mixture immediately prior to compaction shall be within the limits of the compaction temperature established in paragraph (c); otherwise, it shall be discarded. In no case shall the mixture be reheated.

(g) Replace the collar, place the mold assembly on the compaction pedestal in the mold holder. Apply either 35, 50, or 75 blows, as specified according to the design traffic category, with the compaction hammer using a free fall of 18 inches (457 mm). Hold the axis of the compaction hammer as nearly perpendicular to the base of the mold assembly as possible during compaction. Remove the base plate and collar, and reverse and reassemble the mold. Apply the same number of compaction blows to the face of the reversed specimen. After compaction, remove the base plate and allow the specimen to cool in air until no deformation will result when removing it from

Figure III-1—Pedestal, hammer and mold used in preparing Marshall Test specimens

the mold. When more rapid cooling is desired table fans may be used, but not water unless the specimen is in a plastic bag. Remove the specimen from the mold by means of an extrusion jack or other compression device, then place on a smooth, level surface until ready for testing. Normally, specimens are allowed to cool overnight.

C. *Test Procedure*

3.06 GENERAL–In the Marshall method each compacted test specimen is subjected to the following tests and analysis in the order listed:

(a) Bulk Specific Gravity Determination
(b) Stability and Flow Test
(c) Density and Voids Analysis.

3.07 EQUIPMENT–The equipment required for the testing of the 4-inch (102 mm) diameter x 2-1/2-inch (64 mm) height specimens is as follows:

*(a) *Marshall Testing Machine*, an electrically powered (110-volt) testing device. It is designed to apply loads to test specimens through semi-circular testing heads at a constant rate of strain of 2 inches (51 mm) per minute. It is equipped with a calibrated proving ring for determining the applied testing load, a Marshall stability testing head for use in testing the specimen, and a Marshall flow meter for determining the amount of strain at the maximum load for test. A universal testing machine equipped with suitable load and deformation indicating devices may be used instead of the Marshall testing frame.

(b) *Water Bath*, at least 6 inches deep and thermostatically controlled to 140°F ± 1.8°F (60°C ± 1°C). The tank should have a perforated false bottom or be equipped with a shelf for suspending specimens at least 2 inches above the bottom of the bath.

3.08 BULK SPECIFIC GRAVITY DETERMINATION–The bulk specific gravity test may be performed as soon as the freshly compacted specimens have cooled to room temperature. This test is performed according to ASTM Designation D 1188 *Bulk Specific Gravity of Compacted Bituminous Mixtures Using Paraffin-coated Specimens* or ASTM D 2726 *Bulk Specific Gravity of Compacted Bituminous Mixtures Using Saturated Surface-dry Specimens.*

---

*Marshall test apparatus should conform to requirements of ASTM test method D 1559.

3.09 STABILITY AND FLOW TESTS–After the bulk specific gravity of the tests specimen has been determined, the stability and flow tests are performed as follows:

(a) "Zero" the flow meter by inserting a 4.00-inch (101.6 mm) diameter metal cylinder in the testing head, placing the flow meter over the guide rod and adjusting the flow meter to read "zero."

(Note: This adjustment should be made on the guide post marked with an "O" and with the side of the upper segment of the testing head marked with an "O" being placed on the same side as the guide post so marked. The same assembly of testing head and flow meter must then be used in testing the specimens. Specimens should be 4.00 ± 0.01 inches (101.6 ± 0.25 mm); otherwise, an initial and final reading of flow meter is required for the determination of the flow value.)

(b) Immerse specimen in water bath at 140°F ± 1.8°F (60°C ± 1°C) for 30 to 40 minutes before test.

(c) Thoroughly clean inside surfaces of testing head. Temperature of head shall be maintained at from 70° to 100°F (21° to 37.8°C) using a water bath when required. Lubricate guide rods with a thin film of oil so that upper test head will slide freely without binding. If a proving ring is used to measure applied load, check to see that dial indicator is firmly fixed and "zeroed" for the "no-load" position.

(d) With testing apparatus in readiness, remove test specimen from water bath and carefully dry surface. Place specimen in lower testing head and center; then fit upper testing head into position and center complete assembly in loading device. Place flow meter over marked guide rod as noted in (a) above.

(e) Apply testing load to specimen at constant rate of deformation, 2 inches (51 mm) per minute, until failure occurs. The point of failure is defined by the maximum load reading obtained. The total number of pounds (N) required to produce failure of the specimen at 140°F (60°C) shall be recorded as its Marshall stability value.

(f) While stability test is in progress, hold the flow meter firmly in position over guide rod and remove as the load begins to decrease; take reading and record. This reading is the flow value for the specimen, expressed in units of 1/100 inch (0.25 mm). For example, if the specimen deformed 0.15 inch (3.8 mm) the flow value is 15.

(g) The entire procedure, both stability and flow tests, starting with the removal of the specimen from the water bath, shall be completed within a period of thirty seconds.

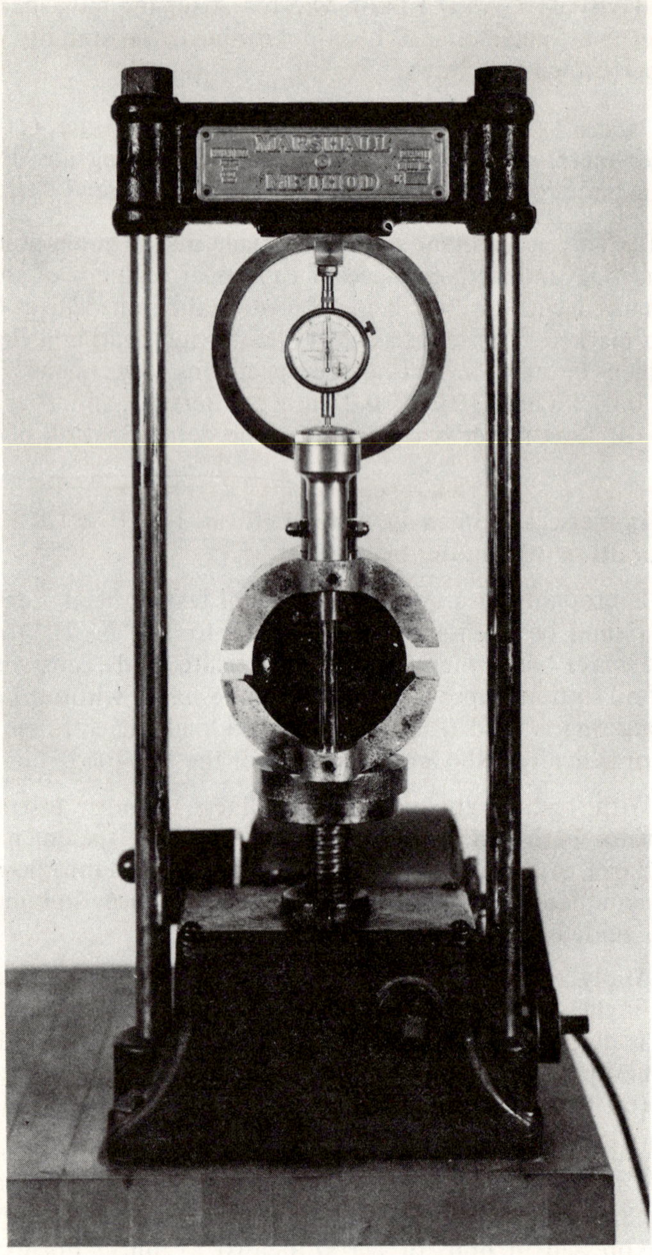

Figure III-2–Marshall stability and flow test

3.10 DENSITY AND VOIDS ANALYSIS–After the completion of the stability and flow tests, a density and voids analysis is made for each series of test specimens as follows:

(a) Average the bulk specific gravity values for all test specimens of a given asphalt content; values obviously in error shall not be included in the average.

(b) Determine average unit weight for each asphalt content by multiplying the average bulk specific gravity value by 62.4 lb/ft^3 (1.0 Mg/m^3) (*See* Figure III-3).

(c) Prepare a graphical plot of unit weight vs. asphalt content and connect plotted points with a smooth curve that obtains the "best-fit" for all values (*See* Figure III-4).

(d) Read unit weight values directly from the plotted curve for each asphalt content tested and compute equivalent bulk specific gravity values by dividing by 62.4 lb/ft^3 (1.0 Mg/m^3). The values of bulk specific gravity thus obtained shall be used in further computations of voids data (*See* Figure III-3).

(e) Using the "bulk" specific gravity of the aggregates and the *Theoretical Maximum Specific Gravity of Bituminous Paving Mixtures* (ASTM D 2041) calculate absorbed asphalt in pounds (kg) of dry aggregate, percent air voids, and percent voids in mineral aggregate (VMA) as illustrated in Figure III-3 and as more fully described in Chapter V. For routine mix design, the *Theoretical Maximum Specific Gravity of Bituminous Paving Mixtures* (ASTM D 2041) is determined in duplicate for at least two asphalt contents, preferably on mixes at or near optimum asphalt content. An average value for asphalt absorption by the aggregate is calculated. This value may then be used for calculation of percent air voids, and percent voids in mineral aggregate, as illustrated in Chapter V. Worksheets for these calculations also are included in Chapter V.

D. *Interpretation of Test Data*

3.11 PREPARATION OF DATA–The stability and flow values and void data are prepared as follows:

(a) Measured stability values for specimens that depart from the standard 2-1/2-inch (63.5 mm) thickness shall be converted to an equivalent 2-1/2-inch (63.5 mm) value by means of a conversion factor. Applicable correlation ratios to convert the measured stability values are set forth in Table III-1. Note that the conversion may be made on the basis of either measured thickness or measured volume.

(b) Average the flow values and the converted stability values for all specimens of a given asphalt content. Values that are obviously in error shall not be included in the average.

HOT MIX DESIGN DATA
by the
MARSHALL METHOD

Trial Mix Series: 3 - B
52% C.A.; 48% F.A.
75 Blow Compaction

Project: F-003-4(2)
Location: Lewis - North
Date: July 17, 1961

Sp. Gr. AC: 1.022
Avg. Bulk Sp. Gr. Total Agg. 2.885

Pen. Grade AC: 85-100
Absorbed AC, lbs./100 lbs. Dry Agg. 0.16

Lab. No. for AC Used: 61 - 205
Lab. Nos. for Agg. Used: 61-551; 61-552; 61-553

% AC-SPEC. NO.	% AC-SPEC. NO.	SPEC. HGT. IN. (cm)	WEIGHT – GRAMS			BULK VOL.-CC.	BULK S.G. COMP.-MIX	AIR VOIDS	% V.M.A.	UNIT WGT. PCF. (Mg/m³)*	STABIL.-LBS (N)**		FLOW 1/100" (0.25 mm)
			IN AIR	IN WATER	SAT. SURFACE DRY IN AIR						MEAS.	ADJUST. †	
% AC by wgt. of Agg.	% AC by wgt. of Mix												
4.17	4.0-A		1306.8	794.8		512.0	2.552				1790	1790	8
	B		1309.5	796.4		513.1	2.563				1810	1810	8
	C		1309.6	798.5		511.1	2.562				1850	1850	8
	Avg.						2.559	5.2	14.8	159.7		1817	8
4.71	4.5-A		1315.3	806.3		509.0	2.584				1900	1900	8
	B		1315.1	806.8		508.3	2.587				1735	1804	10
	C		1311.6	803.5		508.1	2.581				1810	1882	8
	Avg.						2.584	3.5	14.5	161.2		1862	9
5.26	5.0-A		1320.1	812.1		508.0	2.599				1875	1875	11
	B		1318.1	811.0		507.1	2.599				1900	1976	10
	C		1318.7	811.0		507.7	2.597				1740	1810	10
	Avg.						2.598	2.1	14.4	162.1		1887	10
5.82	5.5-A		1320.5	811.5		509.0	2.594				1785	1785	13
	B		1326.1	814.7		511.4	2.593				1685	1685	14
	C		1324.2	814.6		509.6	2.599				1685	1685	13
	Avg.						2.595	1.4	15.0	161.9		1718	13
6.38	6.0-A		1327.6	813.6		514.0	2.583				1490	1490	17
	B		1330.7	815.4		515.3	2.582				1440	1440	17
	C		1329.3	814.4		514.9	2.582				1580	1580	17
	Avg.						2.582	1.1	15.9	161.1		1503	17

*$Mg/m^3 = 0.01618\ lb/ft^3$
**N = 4.448 lbs.

† See Conversion Table III-1

Inspector

Figure III-3–Suggested test report form showing test data for a typical design by the Marshall method

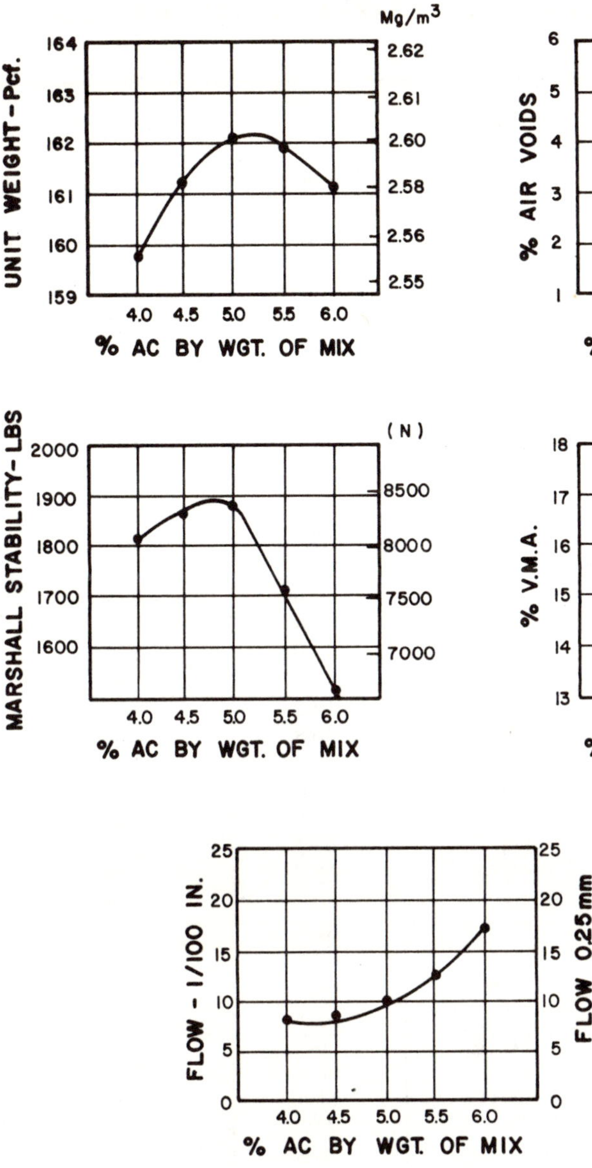

Figure III-4—Test property curves for hot-mix design data by the Marshall method (See also Figure III-3)

TABLE III-1—STABILITY CORRELATION RATIOS

Volume of Specimen, cm^3	Approximate Thickness of Specimen, in.	mm	Correlation Ratio
200 to 213	1	25.4	5.56
214 to 225	1 1/16	27.0	5.00
226 to 237	1 1/8	28.6	4.55
238 to 250	1 3/16	30.2	4.17
251 to 264	1 1/4	31.8	3.85
265 to 276	1 5/16	33.3	3.57
277 to 289	1 3/8	34.9	3.33
290 to 301	1 7/16	36.5	3.03
302 to 316	1 1/2	38.1	2.78
317 to 328	1 9/16	39.7	2.50
329 to 340	1 5/8	41.3	2.27
341 to 353	1 11/16	42.9	2.08
354 to 367	1 3/4	44.4	1.92
368 to 379	1 13/16	46.0	1.79
380 to 392	1 7/8	47.6	1.67
393 to 405	1 15/16	49.2	1.56
406 to 420	2	50.8	1.47
421 to 431	2 1/16	52.4	1.39
432 to 443	2 1/8	54.0	1.32
444 to 456	2 3/16	55.6	1.25
457 to 470	2 1/4	57.2	1.19
471 to 482	2 5/16	58.7	1.14
483 to 495	2 3/8	60.3	1.09
496 to 508	2 7/16	61.9	1.04
509 to 522	2 1/2	63.5	1.00
523 to 535	2 9/16	64.0	0.96
536 to 546	2 5/8	65.1	0.93
547 to 559	2 11/16	66.7	0.89
560 to 573	2 3/4	68.3	0.86
574 to 585	2 13/16	71.4	0.83
586 to 598	2 7/8	73.0	0.81
599 to 610	2 15/16	74.6	0.78
611 to 625	3	76.2	0.76

NOTES:
1. The measured stability of a specimen multiplied by the ratio for the thickness of the specimen equals the corrected stability for a 2 1/2-in. (63.5 mm) specimen.
2. Volume-thickness relationship is based on a specimen diameter of 4 in. (101.6 mm).

(c) Prepare a separate graphical plot for the following values as illustrated in Figure III-4.

Stability vs. Asphalt Content
Flow vs. Asphalt Content
Unit Weight of Total Mix vs. Asphalt Content
Percent Air Voids vs. Asphalt Content
Percent Voids in Mineral Aggregate (VMA) vs. Asphalt Content.

In each graphical plot connect the plotted values with a smooth curve that obtains the "best-fit" for all values.

3.12 TRENDS AND RELATIONS OF TEST DATA—The test property curves, plotted as described above, have been found to follow a reasonably consistent pattern for dense-graded asphaltic paving mixes. Trends generally noted are outlined as follows:

(a) The stability value increases with increasing asphalt content up to a maximum after which the stability decreases.

(b) The flow value increases with increasing asphalt content.

(c) The curve for unit weight of total mix is similar to the stability curve, except that the maximum unit weight normally (but not always) occurs at a slightly higher asphalt content than the maximum stability.

(d) The percent of air voids decreases with increasing asphalt content, ultimately approaching a minimum void content.

(e) The percent voids in the mineral aggregate generally decrease to a minimum value then increase with increasing asphalt contents.

3.13 DETERMINATION OF OPTIMUM ASPHALT CONTENT—The optimum asphalt content of the asphalt paving mix is determined from data obtained as outlined above. Consideration is given to three of the test property curves illustrated in Figure III-4 in making this determination. From these data curves, asphalt contents are determined which yield the following:

(a) Maximum stability

(b) Maximum unit weight

(c) Median of limits given in Article 3.14 for percent air voids.

The optimum asphalt content of the mix is then the numerical average of the values for the asphalt content determined as noted above.

EXAMPLE

Assume that data shown in Figure III-4 represent laboratory tests on a dense-graded asphalt concrete mix [3/4 in. (19 mm) nominal maximum particle size aggregate] to be used for the Heavy traffic category.

Compute optimum asphalt content as follows:

	Percent
(a) Asphalt content at maximum unit weight	5.1
(b) Asphalt content providing 4 percent air voids (median of 3 to 5 percent range for surfacing mix, Heavy and Very Heavy traffic category in the design criteria table below)	4.3
(c) Asphalt content at maximum stability	4.8
Optimum asphalt content, average	4.7

3.14 CRITERIA FOR SATISFACTORY PAVING MIX–Whether the asphalt paving mix will be satisfactory at the optimum asphalt content selected is determined as shown in the above example. This determination is made by applying certain limiting criteria to test data for the mix at its optimum asphalt content. The design criteria in Table III-2 are recommended by The Asphalt Institute.

An example of the application of the above criteria may also be illustrated by use of the data shown in Figure III-4 and information contained in the preceding article.

The optimum asphalt content of the mix, as illustrated, is 4.7 percent. At this asphalt content, properties of the mix are determined from Figure III-4 to be as follows:

(a) Stability	1,880 lb (8,363 N)
(b) Flow	9
(c) Percent air voids	2.8
(d) Percent voids in mineral aggregate	14.4

It will be noted that the stability value exceeds the minimum of 750 pounds (3,336 N), that the flow value is within the limiting range of 8 to 16, and that the percent voids in mineral aggregate exceeds the minimum of 14. The percent voids in the total mix, however, falls below the lower limit shown in the table of criteria for the design conditions previously cited. Normally, adjustments would be made in the mix, as suggested in Article 2.03, to provide a mix design having all tests properties within allowable limits.

3.15 SELECTION OF MIX DESIGN–The selected mix design is usually the most economical one which will satisfactorily meet all of the established criteria. Mixes with abnormally high values of Marshall stability and abnormally low flow values are often less desirable because pavements of such mixes tend to be more rigid or brittle and may crack under heavy volumes of traffic. This is particularly true where base and subgrade deflections are such as to permit moderate to relatively high deflections of the pavement.

TABLE III-2—MARSHALL DESIGN CRITERIA

Traffic Category No. of Compaction Blows Each End of Specimen	Heavy 75		Medium 50		Light 35	
Test Property	Min.	Max.	Min.	Max.	Min.	Max.
Stability, all mixtures, lb (N)	750 (3,336)	–	500 (2,224)	–	500 (2,224)	–
Flow, all mixtures, 0.01 in. (0.25mm)	8	16	8	18	8	20
Percent air voids						
Surfacing or Leveling	3	5	3	5	3	5
Base	3	8	3	8	3	8
Percent voids in mineral aggregate			See Table III-3			

NOTES:
1. Laboratory compactive efforts should closely approach the maximum density obtained in the pavement under traffic.
2. The flow value refers to the point where the load begins to decrease.
3. The portion of the asphalt cement lost by absorption into the aggregate particles must be allowed for when calculating percent air voids.
4. Percent voids in the mineral aggregate is to be calculated on the basis of the ASTM bulk specific gravity for the aggregate.
5. All criteria, and not stability value alone, must be considered in designing an asphalt paving mix.
6. Hot-mix asphalt bases, which do not meet the above criteria when tested at 140°F (60°C), should be satisfactory if they meet the criteria when tested at 100°F (37.8°C) and are placed 4 inches (102 mm) or more below the surface. This recommendation applies only to regions having climatic conditions similar to those prevailing throughout most of the United States. Guidelines for applying the lower test temperature in regions having more extreme climatic conditions are being studied.

In extreme cases when it is not possible or practicable, for economic or other reasons, to meet the requirements of the design criteria, a tolerance of one percent of air voids may be permitted. Under no circumstances, however, shall the allowable flow value be exceeded or shall the stability value be less than that required. It is to be emphasized that these variations should be allowed only under extreme conditions, unless service behavior with a specific aggregate mixture indicates such a paving mix to be satisfactory.

TABLE III-3—MINIMUM PERCENT VOIDS IN MINERAL AGGREGATE (VMA)

U.S.A. Standard Sieve Designation*	Nominal Maximum Particle Size		Minimum Voids in Mineral Aggregate, Percent
	in.*	mm*	
No. 16	0.0469	1.18	23.5
No. 8	0.093	2.36	21
No. 4	0.187	4.75	18
3/8 in.	0.375	9.5	16
1/2 in.	0.500	12.5	15
3/4 in.	0.750	19.0	14
1 in.	1.0	25.0	13
1 1/2 in.	1.5	37.5	12
2 in.	2.0	50	11.5
2 1/2 in.	2.5	63	11

*Standard Specification for Wire Cloth Sieves for Testing Purposes, ASTM Designation E 11 (AASHO Designation M 92).

Chapter IV

HVEEM METHOD OF MIX DESIGN

A. *General*

4.01 DEVELOPMENT AND APPLICATION—The concepts of the Hveem method of designing paving mixtures have been advanced and developed under the direction of Francis N. Hveem, formerly Materials and Research Engineer for the California Division of Highways. The Hveem method has been developed over a period of years during which time certain features have been improved and other features added. The test procedures and their application have been developed through extensive research and correlation studies on asphalt highway pavements.

The Hveem method as developed and used by the California Division of Highways is applicable to paving mixtures using both penetration grades and liquid grades of asphalt and containing aggregates up to 1-inch (25 mm) maximum size. The method shown herein is applicable to the design of hot asphalt paving mixtures. To date, the Hveem method has been used principally for the design of *dense* paving mixtures. The method may be used for both laboratory design and field control of hot-mixed asphalt paving.

Hveem method test procedures have been standardized by the American Society for Testing and Materials. Test procedures are given by ASTM Designation D 1560, *Resistance to Deformation and Cohesion of Bituminous Mixtures by Means of Hveem Apparatus*, and ASTM Designation D 1561, *Compaction of Test Specimens of Bituminous Mixtures by Means of California Kneading Compactor*. Testing procedures presented herein are basically the same as those of the ASTM test methods.

4.02 OUTLINE OF METHOD—The procedure for the Hveem method as presented herein starts with the preparation of test specimens. Preliminary to this operation it is required:

(a) That the materials proposed for use meet the requirements of the project specifications.

(b) That aggregate blend combinations are determined which meet the gradation requirements of the project specifications.

(c) That an ample supply of aggregates is dried and sized into fractions.

These requirements are matters of routine testing, specifications, and laboratory technique, which must be considered but are not unique to any particular mix design method. The reader should refer to Appendix for the schedule of preparation and analysis of aggregates. It should be noted, however, that the maximum size aggregates used in the test mixes should not exceed 1 inch (25 mm). In the event that the specifications for the paving mix being considered require aggregate sizes greater than 1 inch (25 mm), oversize rock up to 25 percent may be screened out without marked effect on stabilometer values.

The Hveem method uses standard test specimens of 2-1/2-inch (64 mm) height by 4-inch (102 mm) diameter; these are prepared using a specified procedure for heating, mixing, and compacting the asphalt-aggregate mixtures. The principal features of the Hveem method of mix design are the Centrifuge Kerosene Equivalent (C.K.E.) test on the aggregates to estimate the asphalt requirements of the mix, followed by a stabilometer test, a cohesiometer test,* a swell test, and a density-voids analysis on test specimens of the compacted paving mixtures. The stabilometer test utilizes a special triaxial-type testing cell for measuring the resistance of the compacted mix to lateral displacement under vertical loading; and the swell test measures the resistance of the mix to the action of water. The specimens are maintained at 140°F (60°C) for the stability tests, whereas the swell tests are performed at room temperature.

B. *Approximate Asphalt Content by the Centrifuge Kerosene Equivalent Method*

4.03 GENERAL–The first step in the Hveem method of mix design is to determine the approximate asphalt content by the Centrifuge Kerosene Equivalent method.† With a calculated surface area and the factors obtained by the C.K.E. method for a particular aggregate or blend of aggregates, the approximate asphalt content is determined by using a series of charts. These charts are presented herein, accompanied by typical examples to demonstrate their application.

---

*The cohesiometer value is seldom used and the test, therefore, is not included in this manual. If the cohesiometer value is desired the test is described in ASTM *Standard Method of Test for Resistance to Deformation and Cohesion of Bituminous Mixtures by Means of Hveem Apparatus*, D 1560

†The development of this method of determining optimum asphalt content is outlined in "Establishing the Oil Content for Dense-Graded Bituminous Mixtures" by F. N. Hveem, *California Highways and Public Works*, July-August, 1942

4.04 EQUIPMENT—The equipment and materials required for determining the approximate asphalt content are as follows:

(a) *Sample Splitter,* small, for obtaining representative samples of fine aggregate.

(b) *Pans,* 4-1/2 inches (115 mm) diameter x 1 inch (25 mm) deep.

(c) *Kerosene,* 1 gallon (4 litres).

(d) *Oil,* SAE No. 10, lubricating, 1 gallon (4 litres).

(e) *Beakers,* 1500 ml.

(f) *Metal Funnels,* 3-1/2 inches (89 mm) top diameter, 4-1/2 inches (114 mm) height, 1/2-inch (13-mm) orifice with piece of 2.00 mm (No. 10) sieve soldered to bottom of opening.

(g) *Timer.*

(h) *Centrifuge,* hand-operated, complete with cups, capable of producing 400 times gravity (Emil Greiner Co., 22 North Moore St., New York, N.Y. 10013, Catalog No. CE-8535 or equal; power-driven centrifuge may be used if available).

(i) *Filter Papers,* 55 mm diam. (No. 611, Eaton-Dikeman Co., Mt. Holly Springs, Pennsylvania, or equivalent).

(Note: See Articles 4.10 and 4.15 for additional equipment requirements.)

Figure IV-1—Apparatus for Hveem C.K.E. tests

4.05 SURFACE AREA–The gradation of the aggregate or blend of aggregates employed in the mix is used to calculate the surface area of the aggregates. This calculation consists of multiplying the total percent passing each sieve size by a "surface-area factor" as set forth in the following tabulation. Add the products thus obtained and the total will represent the equivalent surface area of the sample in terms of square feet per pound (m^2/kg). It is important to note that all surface-area factors must be used in the calculation. Also, if a different series of sieves is used, different surface-area factors are necessary.

TABLE IV-1–SURFACE-AREA FACTORS

Total Percent Passing Sieve No.	Maximum Size	No. 4	No. 8	No. 16	No. 30	No. 50	No. 100	No. 200
Surface-Area Factor,* $ft^2/lb (m^2/kg)$	2 (.41)	2 (.41)	4 (.82)	8 (1.64)	14 (2.87)	30 (6.14)	60 (12.29)	160 (32.77)

*Surface area factors shown are applicable only when all the above-listed sieves are used in the sieve analysis.

The following tabulation demonstrates the calculation of surface area by this method.

Sieve Size	Percent Passing	x	S.A. Factor	=	Surface Area
3/4" (19.0mm)	100		2 (.41)		2.0 (.41)
3/8" (9.52mm)	90				
No. 4	75		2 (.41)		1.5 (.31)
No. 8	60		4 (.82)		2.4 (.49)
No. 16	45		8 (1.64)		3.6 (.74)
No. 30	35		14 (2.87)		4.9 (1.00)
No. 50	25		30 (6.14)		7.5 (1.54)
No. 100	18		60 (12.29)		10.8 (2.21)
No. 200	10		160 (32.77)		16.0 (3.28)

Surface Area = 48.7 $ft.^2/lb$
($9.98 m^2/kg$)

4.06 C.K.E. PROCEDURE–

(a) Place exactly 100 g of dry aggregate (representative of the passing No. 4 material being used) in the tared centrifuge cup assembly fitted with a screen and a disk of filter paper.

(b) Place bottom of centrifuge cup in kerosene until the aggregate becomes saturated.

(c) Centrifuge the saturated sample for two minutes at a force of 400 times gravity. (For the suggested centrifuge this force can be developed by turning the handle approximately 45 revolutions per minute.)

(d) Weigh sample after centrifuging and determine the amount of kerosene retained as a percent of the dry aggregate weight; this value is called the Centrifuge Kerosene Equivalent (C.K.E.). (Note: Duplicate samples are always prepared in order to balance the centrifuge and to check results. The average of the two C.K.E. values is used unless there is a large discrepancy, in which case the test is rerun.)

(e) If the specific gravity of samples is greater than 2.70 or less than 2.60 make a correction to the C.K.E. value using the formula at the bottom of the chart in Figure IV-2.

4.07 SURFACE CAPACITY TEST FOR COARSE AGGREGATE–

(a) Place into a metal funnel, exactly 100 g of dry aggregate passing the 3/8-inch (9.52 mm) sieve and retained on the No. 4 sieve (this fraction is considered to be representative of the coarse aggregate in the mix).

(b) Immerse sample and funnel in a beaker containing SAE No. 10 lubricating oil at room temperature for 5 minutes.

(c) Drain for 2 minutes.

(d) Remove funnel and sample from oil and drain for 15 minutes at a temperature of 140°F (60°C).

(e) Weigh the sample after draining and determine the amount of oil retained as a percent of the dry aggregate weight. (Note: Duplicate samples are prepared to check results. Average value is used unless there is a large discrepancy, in which case the test is rerun.)

(f) If the specific gravity is greater than 2.70 or less than 2.60 make a correction to the percent oil retained using the formula at the bottom of the chart in Figure IV-3.

4.08 ESTIMATED OPTIMUM ASPHALT CONTENT–

(a) Using the C.K.E. value obtained and the chart in Figure IV-2, determine the value K_f (surface constant for fine material).

(b) Using the percent oil retained and the chart in Figure IV-3, determine the value K_c (surface constant for coarse material).

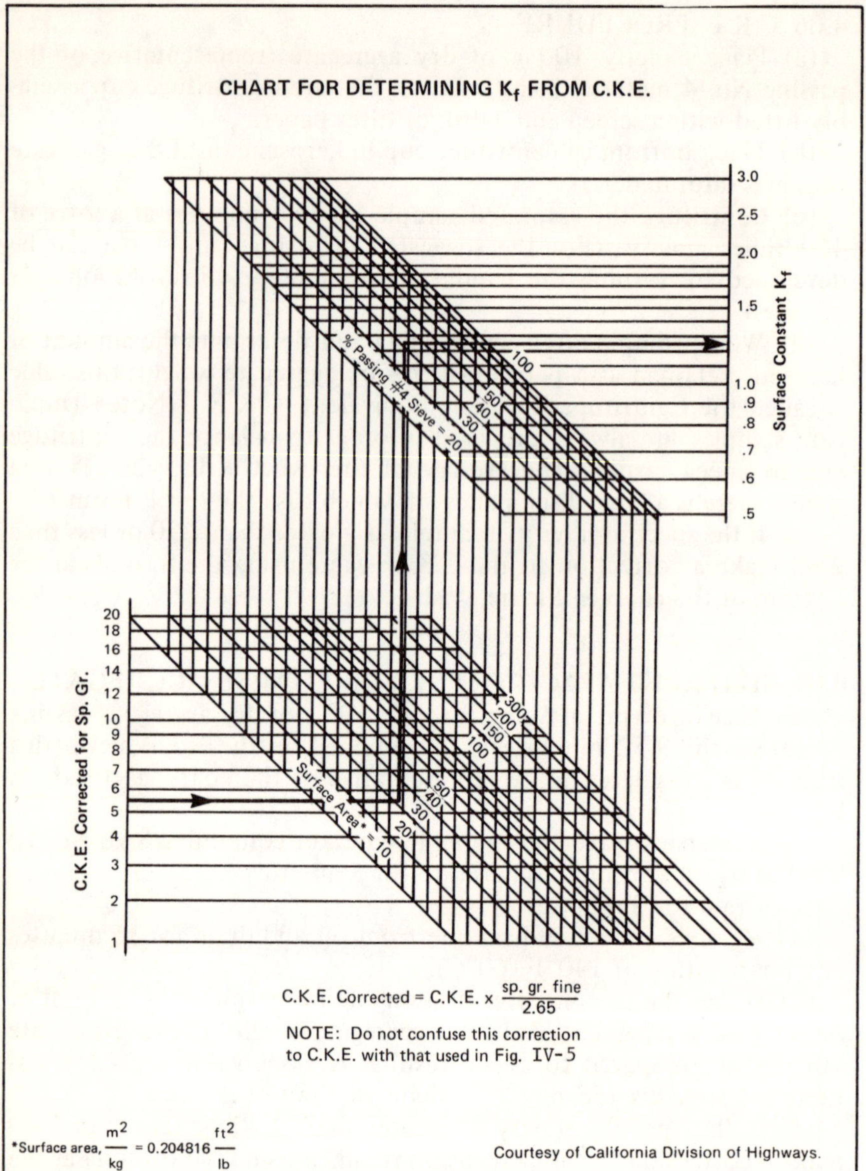

Figure IV-2–Chart for determining surface constant for fine material, K_f, from C.K.E., Hveem method of design

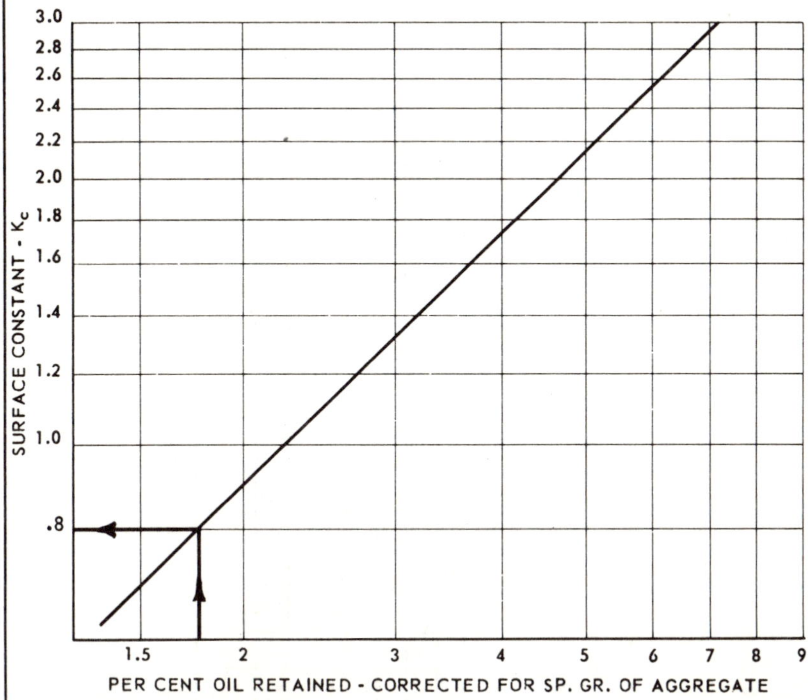

Material Used: Aggregate - Passing 9.5 mm (3/8"), Ret 4.75 mm (#4) Sieve

Oil - SAE 10

$$\% \text{ Oil Ret. Corrected} = \% \text{ Oil Ret.} \times \frac{\text{sp. gr. of Coarse Aggregate}}{2.65}$$

Courtesy of California Division of Highways.

Figure IV-3–Chart for determining surface constant for coarse material, K_c, from coarse aggregate absorption, Hveem method of design

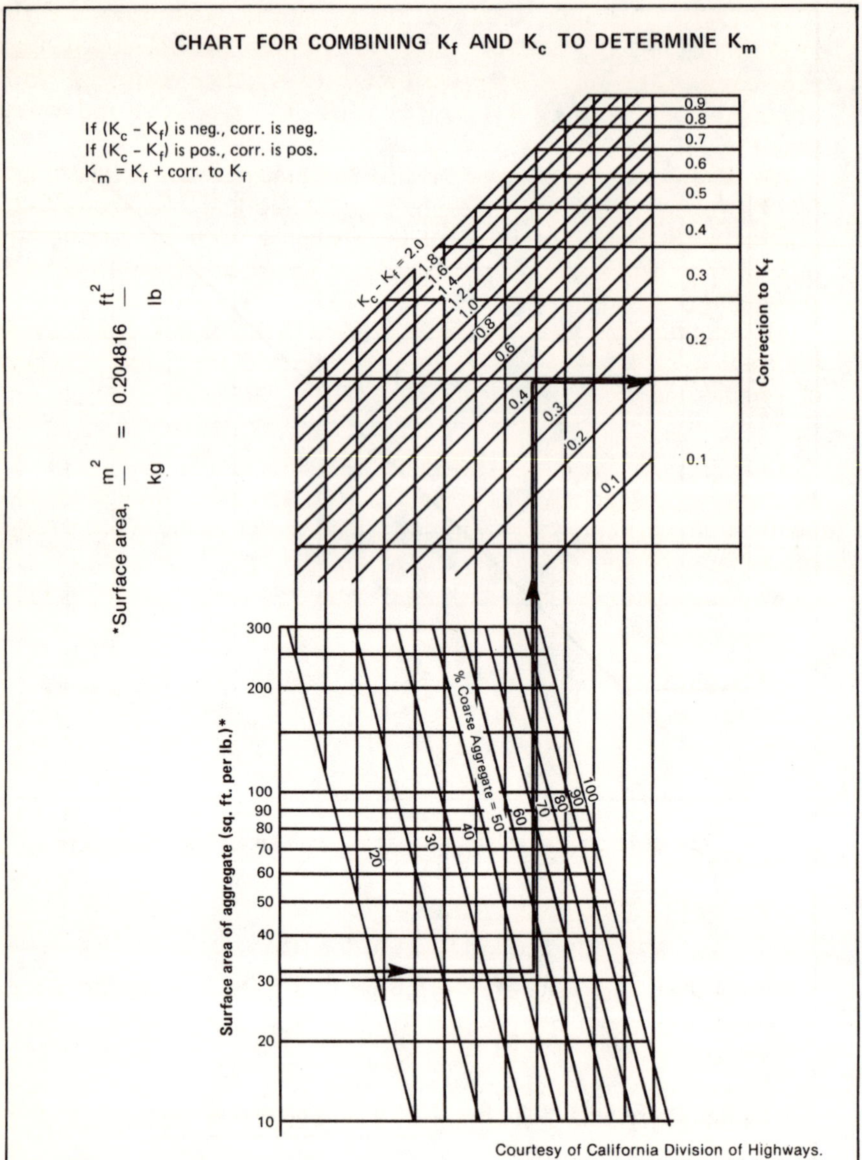

Figure IV-4–Chart for combining K_f and K_c to determine surface constant for combined aggregate, K_m, Hveem method of design

(c) Using the values obtained for K_f and K_c and chart in Figure IV-4, determine the value K_m (surface constant for fine-coarse aggregate combined). $K_m = K_f$ + correction to K_f. The correction to K_f obtained from Figure IV-4 is positive if (K_c-K_f) is positive and is negative if (K_c-K_f) is negative.

(d) The next step is to determine the approximate bitumen ratio for the mix based on cutback asphalts of RC-250, MC-250 and SC-250 grades. With values obtained for K_m, Surface Area and average specific gravity use CASE 2 procedures of Chart in Figure IV-5 to determine the oil ratio.

(e) Determine the asphalt content (Figure IV-6) for mix, corrected for the grade to be employed, using the surface area of the sample, the grade of asphalt, and the oil ratio from Figure IV-5.

EXAMPLE

To demonstrate the use of the charts in Figures IV-2 through IV-6, assume the following conditions apply to a paving mix using AC-10 viscosity grade asphalt cement.

Specific Gravity, coarse = 2.45

Specific Gravity, fine = 2.64

Percent Passing No. 4 = 45

$$\text{Avg. Sp. Gr.} = \frac{100}{\frac{55}{2.45} + \frac{45}{2.64}} = 2.53$$

Surface Area of Aggregate Grading = 32.4 ft^2/lb (6.6 m^2/kg)

C.K.E. = 5.6

Percent Oil Retained, coarse = 1.9

(corrected for specific gravity, this value is 1.7 percent. *See* Figure IV-3)

From Figure IV-2 determine K_f as 1.25.
From Figure IV-3 determine K_c as 0.8.
From Figure IV-4 determine K_m as 1.15.
From Figure IV-5 determine the oil ratio for liquid asphalt as 5.2 percent.
From Figure IV-6 determine optimum asphalt content (bitumen ratio) for AC-10 asphalt as 6.1 percent by weight of dry aggregate.

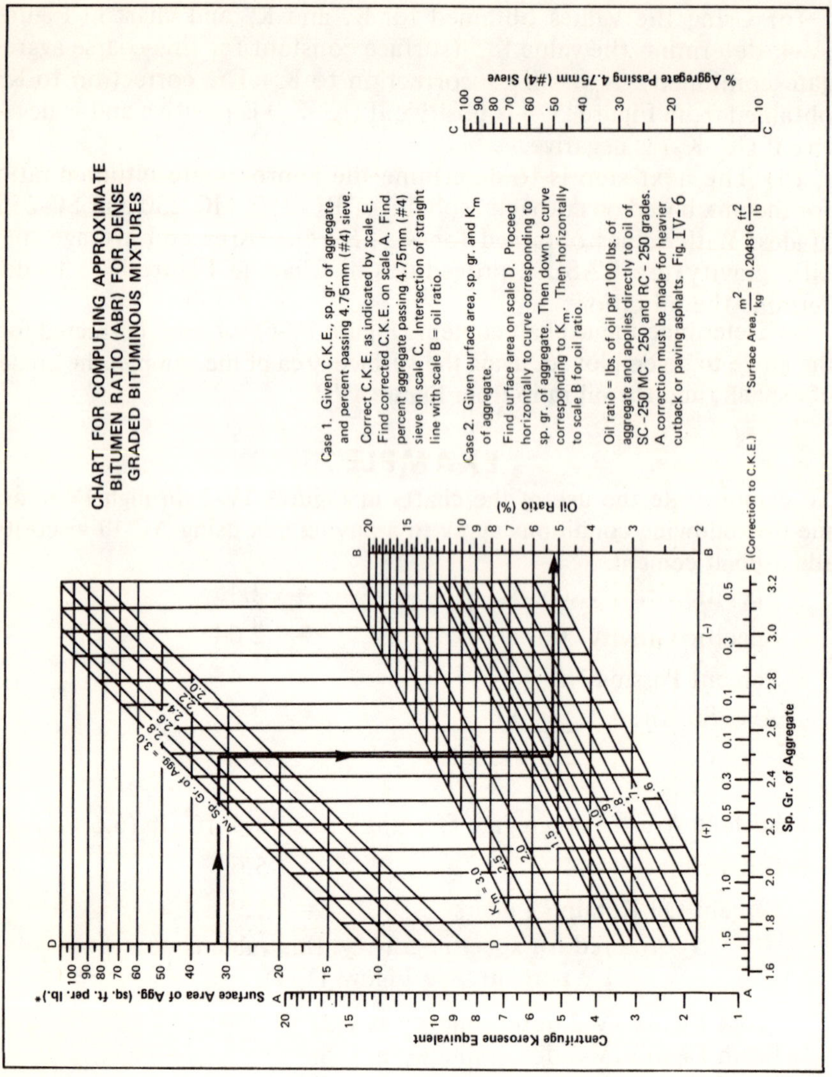

Figure IV-5–Chart for computing oil ratio for dense-graded bituminous mixtures, Hveem method of design

(Chart by courtesy of California Division of Highways)

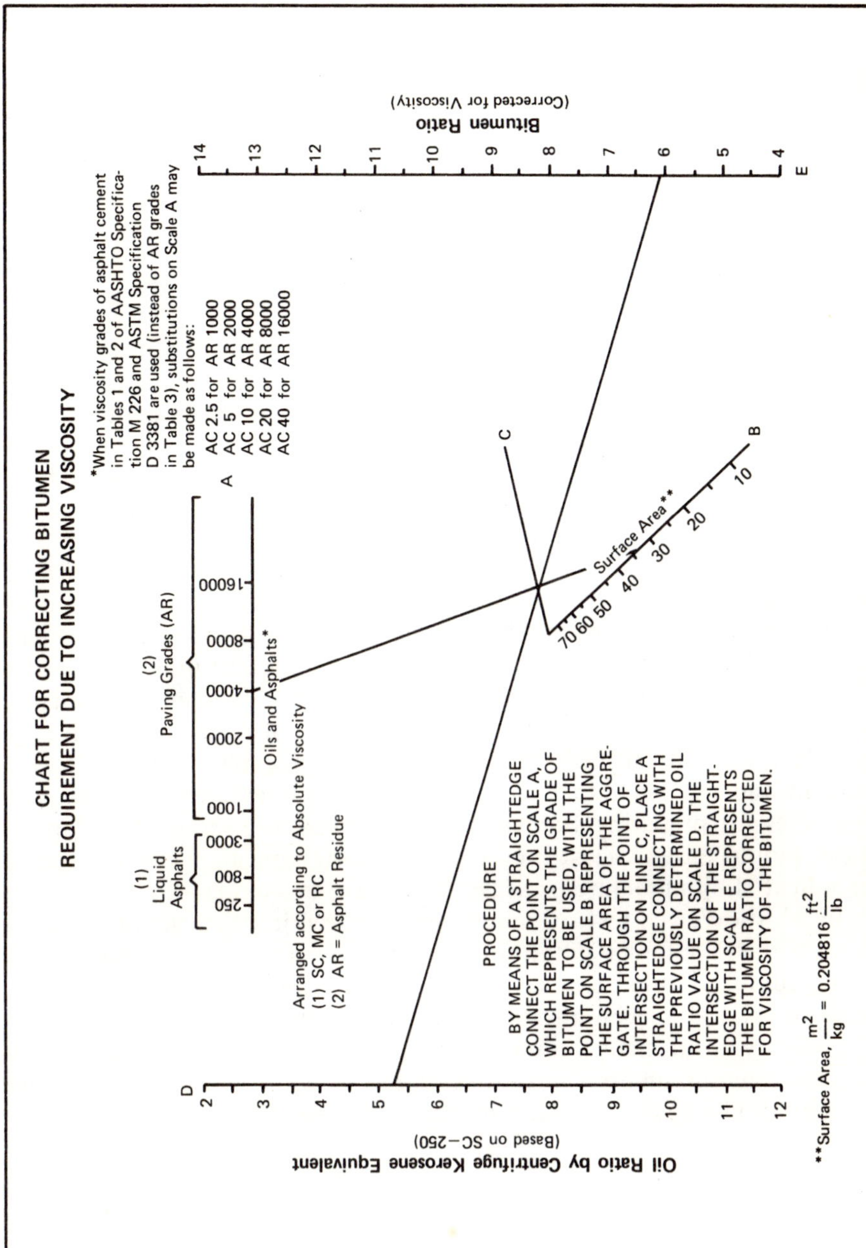

Figure IV-6–Chart for correcting bitumen requirement due to increasing viscosity of asphalt, Hveem method of design
(Chart by courtesy of California Division of Highways)

C. *Preparation of Test Specimens*

4.09 GENERAL–In designing a paving mix by the Hveem method a series of stabilometer test specimens is prepared for a range of asphalt contents both above and below the approximate asphalt content indicated by the CKE procedure.

For hot-mix designs using an average aggregate, tests should be scheduled by preparing one specimen with the amount of asphalt as determined by the CKE procedure, two above the CKE amount in 0.5 increments, and one 0.5 percent below the CKE amount. (Total of four specimens, each with a different asphalt content). For mixes thought to be critical, the steps in asphalt content are lowered to 0.3 percent and tests are scheduled for the approximate asphalt content indicated by the CKE procedure; three above the CKE amount in 0.3 percent increments, and one 0.3 percent below the CKE amount. For highly absorptive aggregates, non-critical mixes increase the steps in asphalt content to 1.0 percent and use more specimens as necessary. Regardless of these general rules for preparing stabilometer specimens, the series of test specimens should have at least one specimen containing an excess of asphalt as indicated by moderate or heavy flushing after compaction.

In addition swell test specimens are prepared in duplicate at the optimum asphalt content as determined from tests on the series of specimens prepared for stabilometer tests. Thus for a normal mix design study a total of six test specimens will usually be required. Although each test specimen will normally require only 1.2 kg of aggregate, the minimum aggregate requirements for a series of test specimens should be at least 40 pounds (18 kg) to provide for additional tests that may be required.

4.10 EQUIPMENT–The equipment required for the preparation of test specimens is as follows:

(a) *Pans,* 10 inches (250 mm) diameter x 2 inches (50 mm) deep, for quartering and mixing fine aggregate.

(b) *Pans,* 8 inches (200 mm) diameter x 1-3/4 inches (45 mm) deep, for batching and heating aggregates.

(c) *Pans,* 12 inches (305 mm) diameter x 2-1/2 inches (64 mm) deep for mixing aggregate and asphalt.

(d) *Pans,* 11 inches (280 mm) x 7 inches (180 mm) x 1-1/2 inches (40 mm) for curing mix.

(e) *Sample Splitter,* large, for mixing and quartering fine aggregate.

(f) *Hot Plate,* electric, at least 18 inches (460 mm) x 12 inches (305 mm), plate surface, for heating aggregates, asphalt and equipment as required.

(g) *Oven,* large, thermostatically controlled, capable of 230°F (110°C) temperature.

(h) *Oven,* large, thermostatically controlled, capable of 140°F (60°C) temperature.

(i) *Oven,* large, for drying and preheating, capable of temperatures up to 325°F (163°C).

(j) *Scoop,* large, for handling hot aggregates.

(k) *Beakers,* 800 ml, for adding asphalt.

(l) *Thermometer,* armored, 100°F (38°C) to 400°F (204°C).

(m) *Balance,* 20-kg capacity, sensitive to 1.0 g for weighing aggregates and asphalt.

(n) *Mixing Trowel,* small pointed.

(o) *Mixing Spoon,* large.

(p) *Mechanical Mixer* (optional).

(q) *Mechanical Compactor* (*see* Figure IV-7) designed to consolidate the material by a series of individual "kneading action" impressions made by a roving ram having a face shaped as a sector of a 4 inch (101.6 mm) diameter circle (*see* Figure IV-8). The compactor must be capable of exerting a force of 500 psi (34.5 kPa) under the tamper foot. Accessories with the compactor should include two mold holders, a feeder trough 18 inches (460 mm) long x 4 inches (102 mm) wide x 2-1/2 inches (64 mm) deep and a paddle shaped to **fit trough, and a bullet-nosed steel rod 3/8-inch (9.5 mm) diameter x** 16 inches (405 mm) long.

(r) *Compaction Molds,* steel, 4 inches (101.6 mm) inside diameter x 5 inches (127 mm) high x 1/4-inch (6.4 mm) wall thickness.

(s) *Paper Disks,* heavy paper, 3-15/16 inches (100 mm) diameter, to place in bottom of mold during compaction.

(t) *Compression Machine,* hydraulic 50,000-pound (222.4 kN) capacity.

(u) *Gloves,* leather or asbestos, for handling hot equipment.

(Note: See Articles 4.04 and 4.15 for additional equipment requirements.)

Figure IV-7–Mechanical kneading compactor for the preparation of test specimens

4.11 BATCH WEIGHTS–

(a) Compute batch weights for the blend and gradation of aggregates desired. Suggested procedures for computing batch weights are presented in the Appendix.

(b) The dry weight of the aggregate for the swell test specimen shall always be 1000 g. Normally 1200 g (dry weight) is sufficient for the stabilometer specimen. However, if the average specific gravity of the aggregate is 2.80 or higher, 1300 g (dry weight) will be required.

4.12 PREPARATION OF BATCH MIXES–

(a) Weigh the various-sized fractions of dry aggregates into suitable pans in accordance with accumulative batch weights. (*See* Appendix.)

(b) Thoroughly mix each individual batch of aggregate and place in oven to preheat to desired mixing temperature. Asphalt should be preheated at the same time. The temperature of the aggregate and asphalt at the time mixing begins is indicated below for the paving grade of asphalt cement being used.

Grade	Temperature Range Minimum	Maximum
AC-2.5, AR-1000, or 200-300 Pen.	225°F (107°C)	250°F (121°C)
AC-5, AR-2000, or 120-150 Pen.	250°F (121°C)	275°F (135°C)
AC-10, AR-4000, or 85-100 Pen.	275°F (135°C)	300°F (149°C)
AC-20, AR-8000, or 60-70 Pen.	300°F (149°C)	325°F (163°C)
AC-40, AR-16000, or 40-50 Pen.	300°F (149°C)	325°F (163°C)

(c) When the aggregates and asphalt have reached the mixing temperature indicated above, form a crater in the aggregates and weigh in asphalt in accordance with accumulative batch weights.

(d) Place pan containing aggregates and asphalt for batch mix on hot plate to maintain mixing temperature. Vigorously hand-mix aggregates and asphalt with pointed trowel until all particles are coated. Take special precaution not to overheat the materials. The mixing operation may also be accomplished by mechanical mixing.

(e) After mixing is complete, transfer the batch mix to a suitable flat pan and place in oven for a 15-hour curing period at 140°F ± 5°F (60°C ± 2.8°C). The oven used for curing should preferably be equipped for forced draft air circulation.

(f) After curing is complete, place batch mix in heating oven and reheat mixture to 230°F (110°C). The batch mix is then ready for compaction.

4.13 COMPACTION–The compaction of the test specimen is accomplished by means of the mechanical compactor that imparts a kneading action consolidation by a series of individual impressions made with a ram having a face shaped as a sector of a 4 inch (101.6

mm) diameter circle. At each application of the ram a pressure of 500 psi (3.45 MPa) is applied, subjecting the specimen to a kneading action without impact over an area of approximately 3.1 square inches (20 cm^2). Each pressure is maintained for approximately 2/5ths of a second. The detailed compaction procedure is as follows:

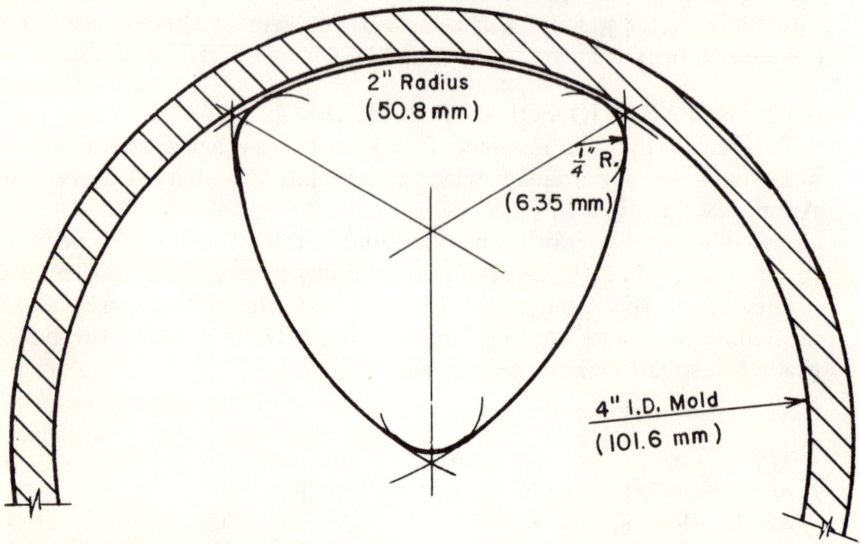

Figure IV-8—Diagram of tamping foot for mechanical kneading compactor

For Swell Test Specimens

(a) Prepare compaction mold by placing a paraffin-impregnated strip of ordinary wrapping paper 3/4-inch (19 mm) wide, around the inside of mold 1/2-inch (13 mm) to 3/4-inch (19 mm) from the bottom to prevent water from escaping from between the specimen and the mold during the water immersion period of the test. The paper strip is dipped in melted paraffin and applied while hot. Compaction molds are not preheated for swell test specimens.

(b) Heat compactor foot to a temperature that will prevent the mix from adhering to it. Temperature of the compactor foot may be controlled by a variable transformer.

(c) Place the compaction mold in position in the mold holder with a 4 inch (101.6 mm) diameter paper disc inserted to cover the base plate. In order to have the base plate act as a free-fitting plunger during the compaction operation, a steel shim 1/4-inch (6.4 mm) thick x 3/4-inch (19.0 mm) wide x 2-1/2 inches (63.5 mm) long is temporarily placed under the edge of the mold.

(d) Weigh 1000 ± 1 g of mixture, heated to the compaction temperature of 230°F (110°C) into the insulated feeder trough. The trough should be preheated to approximately 230°F (110°C).

(e) Spread mixture uniformly in feeder trough. With a paddle made to fit the shape of the trough, transfer approximately one-half of the mixture to the compaction mold (*See* Figure IV-9).

(f) Rod the portion of the mix in the mold 20 times in the center of the mass and 20 times around the edge with a bullet-nosed steel rod, 3/8-inch (9.5 mm) diameter x 16 inches (405 mm) long. Transfer the remainder of the sample to the mold and repeat the rodding procedure (*See* Figure IV-10).

Figure IV-9–Transfer of mix to mold

Figure IV-10–Rodding mix in mold

(g) Place mold assembly into position on the mechanical compactor and apply approximately 20 tamping blows at 250 psi (1.7 MPa) pressure to accomplish a semi-compacted condition of the mix so that it will not be unduly disturbed when the full load is applied. The exact number of blows to accomplish the semi-compaction shall be determined by observation. The number of blows may vary between 10 and 50, depending upon the type of material, and in some instances it may not be possible to accomplish the compaction in the mechanical compactor because of undue movement of the mixture under the compactor foot. In these instances use a 40,000-pound (177.93 kN) static load applied by the double plunger method, in which a free-fitting plunger is placed below the sample as well as on top. Apply the load at the rate of 0.05 in. (1.3 mm) per minute and hold for 30 ± 5 seconds.

(h) After the semi-compaction, remove shim and release mold tightening screw sufficiently to allow free up-and-down movement of mold.

(i) To complete compaction in the mechanical compactor, increase compactor foot pressure to 500 psi (3.45 MPa) and apply 150 tamping blows.

(j) Remove mold and specimen from compactor, invert mold and push specimen to the opposite end of mold. Apply a 1000 psi (6.9 MPa) static load [head speed 0.25 in. per minute (6.3 mm/min.)] with the "original" top surface supported on the lower platen of testing press. It is advisable to place a piece of heavy paper under the specimen to prevent damage to this lower platen. Remove the specimen from the mold. (ASTM Method D 1560 describes a push-out device for removing specimens.) Measure and record the specimen height.

For Stabilometer Specimens

(a) Preheat compaction molds to approximately the temperature to be used for the compaction of the mix.

(b) Using the swell test specimens as pilot samples, adjust the amount of mix as follows to obtain a 2.5-inch (63.5 mm) compacted height for the stabilometer cohesiometer specimens.

Adjusted Weight of Mix, g =

$$\frac{2.5 \text{ in.} \times 1000}{\text{Swell Test Specimen Height, in.}}$$

or, for International System of Units (SI),

Adjusted Weight of Mix, g =

$$\frac{63.5 \text{ mm} \times 1000}{\text{Swell Test Specimen Height, mm}}$$

(c) The remainder of the compaction procedure for stabilometer specimens is the same as for the swell test specimens with the following exception:

> Paraffin strips are not used. When compaction is completed in the mechanical compactor, the mold and specimen are placed in the

140°F (60°C) oven for 1-1/2 hours, after which a "leveling-off" load of 1000 psi (6.9 MPa) is applied by the "double-plunger" method [head speed 0.25 inch per minute (6.3 mm/min.)] and released immediately. (Note: The specimen shall not be pushed to the opposite end of the mold.)

D. *Test Procedures*

4.14 GENERAL–In the Hveem method the compacted test specimens are used in the following tests and analyses and are normally performed in the order listed:

(a) Stabilometer Test
(b) Bulk Density Determination
(c) Swell Test.

The swell test is performed only on specimens prepared for this purpose; the stabilometer and bulk density tests are performed on each of all other test specimens. Figure IV-15 shows a suggested test report form for recording test data and results.

4.15 EQUIPMENT–The equipment required for the testing of the 4 inch (102 mm) diameter specimens is as follows:

(a) *Bronze Disks,* perforated, 3-7/8-inch (98.4 mm) diameter x 1/8-inch (3.2 mm) thick, with adjustable stem, for swell measurement. (*See* Figure IV-11).

(b) *Dial Gauge,* mounted on tripod, with reading accuracy to 0.001 inch (0.025 mm) (*See* Figure IV-11).

(c) *Scale,* graduated to read the volumetric contents of a 4 inch (101.6 mm) inside-diameter mold at 25 ml intervals, for measuring percolation of water during swell test.

(d) *Pans,* aluminum, 7-1/2-inch (190 mm) diameter x 2-1/2-inches (64 mm) deep.

(e) *Hveem Stabilometer* (*see* Figures IV-12 and IV-13), complete with accessories including adjustable base, assembly tool, steel follower, and rubber bulb for introducing air into system.

(f) *Scale* or other measuring device to determine accurately height of compacted test specimen.

4.16 STABILOMETER TEST *(see* Figures IV-12 and IV-13)–

(a) Place specimens for stabilometer tests (compacted and contained in mold) in oven at 140°F (60°C) for a minimum period of one hour before testing.

(b) Adjust compression machine for a head speed of 0.05 inch per minute (0.02 mm/s) with no load applied.

(c) Check displacement of stabilometer with metal dummy specimen and if necessary adjust to give 2.00 ± 0.05 turns (see Article 4.21).

(d) Adjust the stabilometer base so that the distance from the bottom of the upper tapered ring to the top of the base is 3.5 in. (89 mm).

(e) Every effort should be made to fabricate test specimens with an overall height between 2.40 inches (61 mm) and 2.60 inches (66 mm); however, if the height is outside of this range the stabilometer value should be corrected as indicated in Figure IV-14.

(f) Remove mold containing specimen from oven and place on top of stabilometer. Using the plunger, hand lever, and fulcrum, force the specimen from the mold into the stabilometer. Take care that the specimen goes in straight and is firmly seated.

(g) Place follower on top of specimen and position the entire assembly in compression machine for testing.

(h) Using a displacement pump, raise the pressure in the stabilometer system until test gauge reads exactly 5 psi (34.5 kPa). Then tap test gauge lightly to assure an accurate reading.

(i) Close displacement pump valve, taking care not to disturb the 5 psi (34.5 kPa) initial pressure. (This step is omitted on stabilometers that are not provided with the displacement pump valve.)

(j) Apply test loads with compression machine using a head speed of 0.05 inch per minute (0.02 mm/s). Record readings of test gauge at test loads of 500 lb (2.22 kN), 1,000 lb (4.45 kN) and each 1,000 lb (4.45 kN) thereafter up to a maximum of 6,000 lb (26.69 kN).

(k) Immediately after recording the reading under maximum load [6,000 lb (26.69 kN)], reduce total load on specimen to 1,000 lb (4.45 kN).

(l) Open the displacement pump angle valve and by means of the displacement pump, adjust test gauge to 5 psi (34.5 kPa) (this will result in a reduction in the applied press load which is normal and for which no compensation is made).

(m) Adjust dial gauge on pump to zero by means of small thumbscrew.

(n) Turn displacement pump handle rapidly (two turns per second) and smoothly to right (clockwise) until a pressure of 100 psi (690 kPa) is recorded on test gauge. (During this operation the load registered on the testing press will increase and in some cases exceed the initial 1,000 lb (4.45 kN) load. This change in load is normal and no adjustment or compensation is required under the present procedure.) Record the exact number of turns required to increase the test gauge reading from 5 psi (34.5 kPa) to 100 psi (690 kPa) as *displacement* on specimen. [0.1 in. (2.5 mm) dial reading is equivalent to one turn displacement.]

(o) After recording the displacement, first remove the test load and reduce pressure on test gauge to zero by means of displacement pump; then back off displacement pump an additional three turns and remove specimen from stabilometer chamber.

4.17 BULK DENSITY DETERMINATION–The bulk density test is performed on the specimens after the completion of the stabilometer tests and as soon as the specimens have cooled to room temperature. The procedure for this test is presented in ASTM Designation D 1188, *Bulk Specific Gravity of Compacted Bituminous Mixtures Using Paraffin-Coated Specimens* or ASTM Designation D 2726, *Bulk Specific Gravity of Compacted Bituminous Mixtures Using Saturated Surface-Dry Specimens.*

4.18 SWELL TEST–

(a) Allow compacted specimen for swell test to stand at room temperature for at least one hour (this is done to permit rebound after compaction).

(b) Place mold and specimen in 7-1/2-inch (190 mm) diameter x 2-1/2-inch (64 mm) deep aluminum pan. (*See* Figure IV-11.)

(c) Place perforated bronze disk on specimen, position tripod with dial gauge on mold, and set adjustable stem to give a reading of 0.100 inch (2.54 mm) on dial gauge. (*See* Figure IV-11.)

(d) Introduce 500 ml of water to mold on top of specimen and measure distance from top of mold to surface of water with graduated scale.

(e) After 24 hours again read dial gauge and record the change as *swell* to nearest 0.001 inch (0.025 mm); also, again measure distance from top of mold to surface of water with graduated scale and record the change as *permeability* or the amount of water in milliliters that percolates into and/or through the test specimen.

E. *Interpretation of Test Data*

4.19 CALCULATIONS–There are no calculations required for the swell test since the results are reported directly as differences. The remainder of the calculations are as follows:

(a) *Stabilometer Value.* Calculate as:

$$S = \frac{22.2}{\frac{P_h D_2}{P_v - P_h} + 0.222}$$

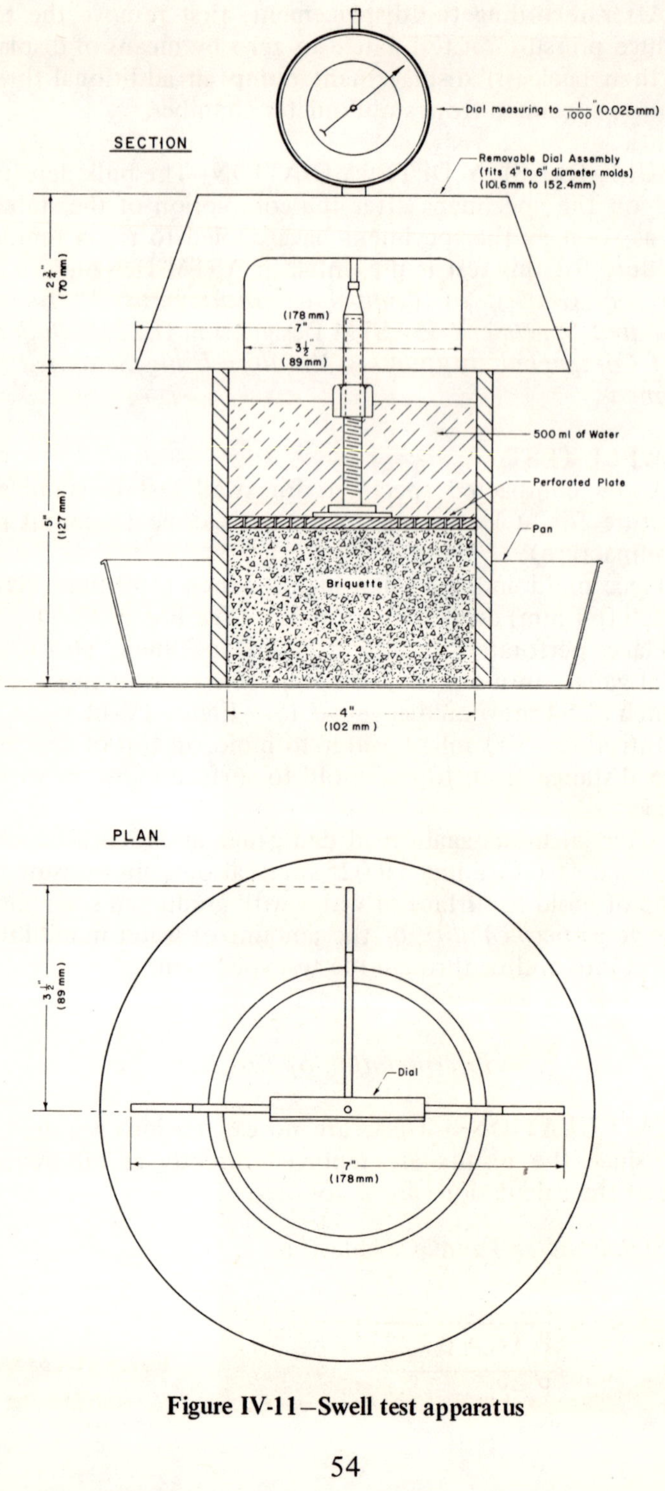

Figure IV-11–Swell test apparatus

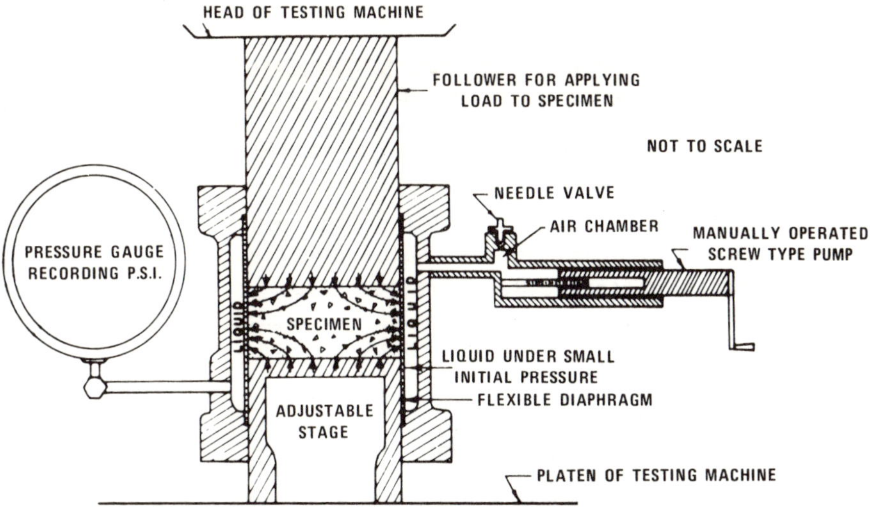

NOTE–The specimen is given lateral support by the flexible sidewall, which transmits horizontal pressure to the liquid. The magnitude of the pressure can be read on the gauge.

Figure IV-12–Diagrammatic sketch showing principal features of Hveem stabilometer

Figure IV-13–Hveem stabilometer

CHART FOR CORRECTING STABILOMETER VALUES TO SPECIMEN HEIGHT OF 2.50

Height correction should be made using the table and chart below.

Example: Overall height of 2.74", select correction curve "B". Stabilometer value uncorrected = 35 Stabilometer value corrected = 38.

Overall Specimen Ht.	Correction Curve
2.80" to 3.00"	A
2.60" to 2.79"	B
2.40" to 2.59"	C
2.20" to 2.39"	D
2.00" to 2.19"	E

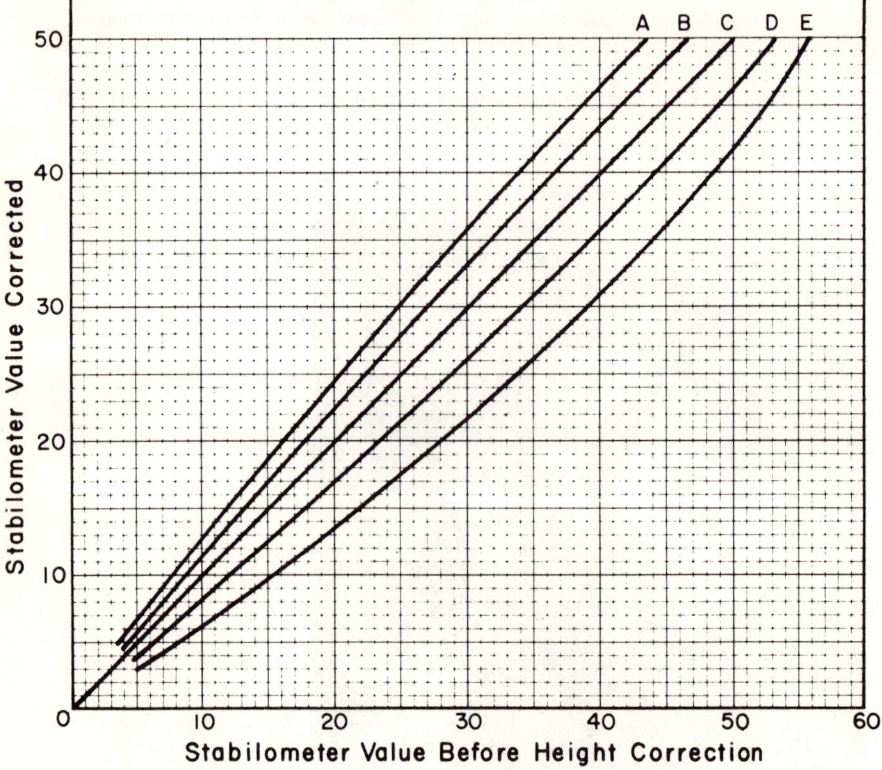

NOTE : TO CONVERT FROM INCHES TO MILLIMETERS MULTIPLY BY 25.4

Figure IV-14–Chart for correcting stabilometer values to effective specimen height of 2.40 inches (61 mm)

where: S = stabilometer value
D_2 = displacement on specimen
P_v = vertical pressure [typically 400 psi (2.76 MPa) = 5000 lb (22.24 kN) total load]
P_h = horizontal pressure = stabilometer pressure gauge reading taken at the instant P_v is 400 psi (2.76 MPa) [5,000 lb (22.24 kN) total load]

(b) *Density and Voids Analysis.* Using the bulk specific gravity of the test specimens, maximum specific gravity of the paving mixture, the "apparent" specific gravity of aggregate, and the specific gravity of other constituent materials in the mix determined as noted in Article 2.08, compute the percent air voids as illustrated in Figure IV-15 and more fully described in Chapter V.

4.20 DESIGN CRITERIA—The suitability of the hot-mix design by the Hveem method is determined on the basis of whether the asphalt content and aggregate grading will satisfy the following requirements:

TABLE IV-2—HVEEM DESIGN CRITERIA

Traffic Category	Heavy		Medium		Light	
Test Property	min.	max.	min.	max.	min.	max.
Stabilometer Value	37	–	35	–	30	–
Swell	less than 0.030 inch (0.762 mm)					

NOTES:
1. Although not a routine part of this design method, an effort is made to provide a minimum percent air voids of approximately 4 percent.
2. All criteria, and not stability value alone, must be considered in designing an asphalt paving mix.
3. Hot-mix asphalt bases, which do not meet the above criteria when tested at 140°F (60°C) should be satisfactory if they meet the criteria when tested at 100°F (38°C) and are placed 4 inches (102 mm) or more below the surface. This recommendation applies only to regions having climatic conditions similar to those prevailing throughout most of the United States. Guidelines for applying the lower test temperature in regions having more extreme climatic conditions are being studied.

In the application of these requirements, the optimum asphalt content for the mix design should be the highest percentage the mix will accommodate without reducing stability or void content below minimum values. The optimum asphalt content is determined from stabilometer

Trial Mix Series: 1-B
53% CA; 47% FA

HOT-MIX DESIGN DATA by the HVEEM METHOD

Project: F1-008-8(3)
Location: Rye - South
Date: August 4, 1973

Sp. Gr. Asp. Cem. 1.012 Asp. Cem. AC-10 — Lab. No. for Asp. Cem. Used: 53-0741
Avg. Bulk Sp. Gr. Agg. = 2.760 — Lab. Nos. for Agg. Used: 53-1252; 53-1253

GRADATION, CKE, and PERCENT ASPHALT

Sieve Size	1½	1	¾	½	3/8	4	8	16	30	50	100	200
Specification Limits			100	100/80	90/70	70/50	50/35	/	29/18	/	16/8	10/4
% Passing			100	91	76	60	42	32	23	16	12	6
S. A. Factors						2	4	8	14	30	60	160
Surface Area, ft^2/lb*					2	1.2	1.7	2.6	3.2	4.8	7.2	9.6

CKE: FA = 2.8; CA = 2.8 K_f = 1.0; K_c = 1.3; K_m = 1.0; Total SA 32.3 Sq. ft^2/lb. (6.62 m^2/kg)

Estimated % Asp. Cem. by Wgt. of Agg. using CKE Tests only 5.5

Recommended % Asp. Cem. by Wgt. of Agg. using Mix Design Criteria 5.0

Specimen Identification	A	B	C	D
% Asp. Cem. by Wgt. of Agg.	5.0	5.5	6.0	6.5
% Asp. Cem. by Wgt. of Mix	4.76	5.21	5.66	6.10
Wgt. in Air-grams	1211.0	1223.3	1230.8	1235.9
Wgt. in Water-grams	714.9	723.8	727.6	733.3
Bulk Volume-cc.	496.1	499.5	503.2	502.6
Bulk Sp. Gr.	2.441	2.449	2.446	2.459
Max Sp. Gr.	2.559	2.540	2.522	2.504
% Voids-Total Mix	4.6	3.6	3.0	1.8
Unit Wgt. - pcf. (kg/m^3)	152.3(2.439)	152.8(2.448)	152.6(2.446)	153.4(2.457)

Total Load-lbs.	(kN)	Unit Load-psi	(MPa)	STABILOMETER A	B	C	D
500	(2.22)	40	(0.28)	9	9	9	10
1000	(4.45)	80	(0.55)	12	12	15	16
2000	(8.90)	160	(1.10)	15	16	24	26
3000	(13.34)	240	(1.65)	21	22	30	38
4000	(17.79)	320	(2.21)	28	30	42	55
5000	(22.24)	400	(2.76)	36	39	55	83
6000	(26.69)	480	(3.31)	50	52	62	105
Displacement-turns				2.40	2.50	2.46	2.50
Stability Value				48	45	36	25

COHESIOMETER

		A	B	C	D
Temperature - °F. Effective Height-in. (cm) Shot Weight - grams Cohesiometer Value					

Jones
Inspector

*Surface Area, m^2/kg = 0.204816 ft^2/lb

Figure IV-15–Suggested test report form showing test data for a typical mix design by the Hveem method

values, percent air voids, and observations of surface flushing or bleeding of specimens after compaction. The following steps are used to select the optimum asphalt content:

(a) Using Figure IV-17, Insert in Step (1) of the pyramid, the asphalt contents used for preparing the series of mix design specimens. Insert asphalt contents in order of increasing amounts from left to right with the maximum asphalt content used in the square on the right.

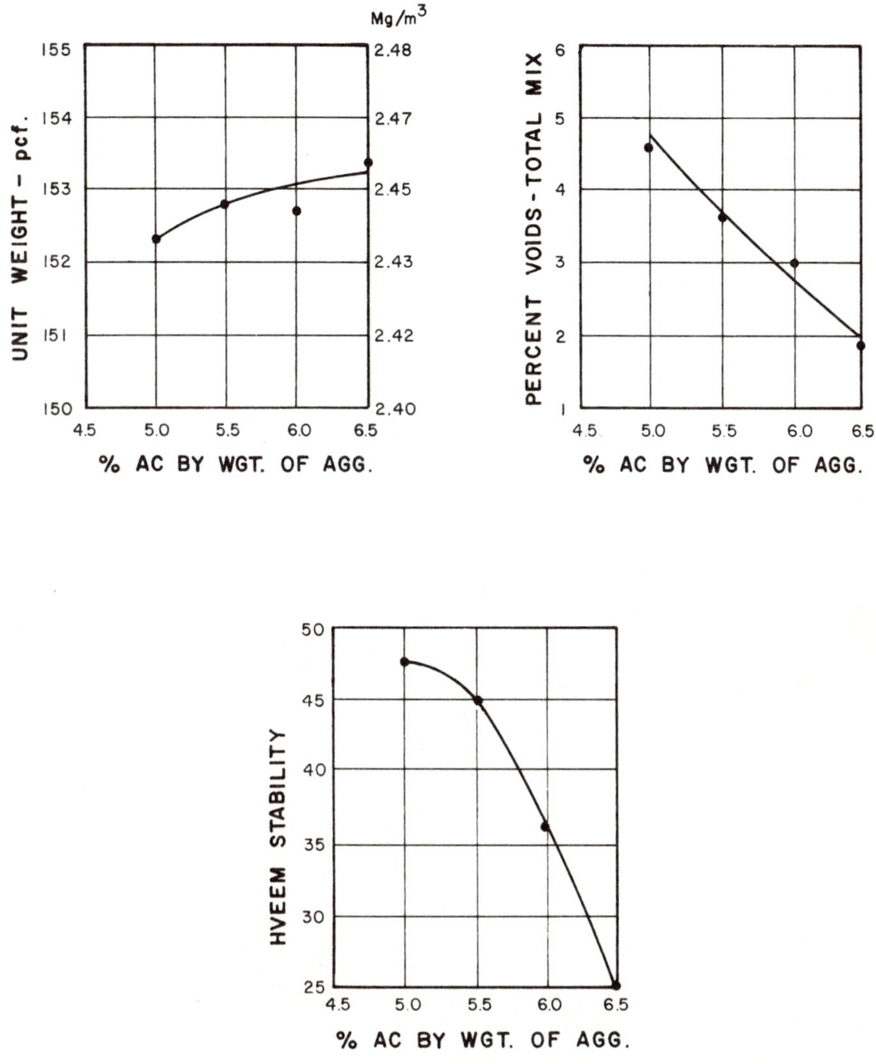

Figure IV-16–Test property curves for hot-mix design data by the Hveem method (see also Figure IV-15)

(b) Select from Step (1) the three highest asphalt contents that do not exhibit moderate or heavy surface flushing and record on Step (2). (Surface flushing and/or bleeding is considered "Slight" if the surface has a slight sheen. It is considered "Moderate," if sufficient free asphalt is apparent to cause paper to stick to the surface but no distortion is noted. Surface flushing is considered "Heavy" if there is sufficient free asphalt to cause surface puddling or specimen distortion after compaction).

(c) Select from Step (2) the two highest asphalt contents that provide the specified minimum stabilometer value and enter them in Step (3).

(d) Select from Step (3) the highest asphalt content that has at least 4.0% air voids and enter in Step (4).

(e) The asphalt content in Step (4) is the optimum asphalt content. However, if the maximum asphalt content used in the design set (Step 1) is the asphalt content entered on Step (4), additional specimen must be prepared with increased asphalt content in 0.5 percent increments and a new optimum asphalt content determination made.

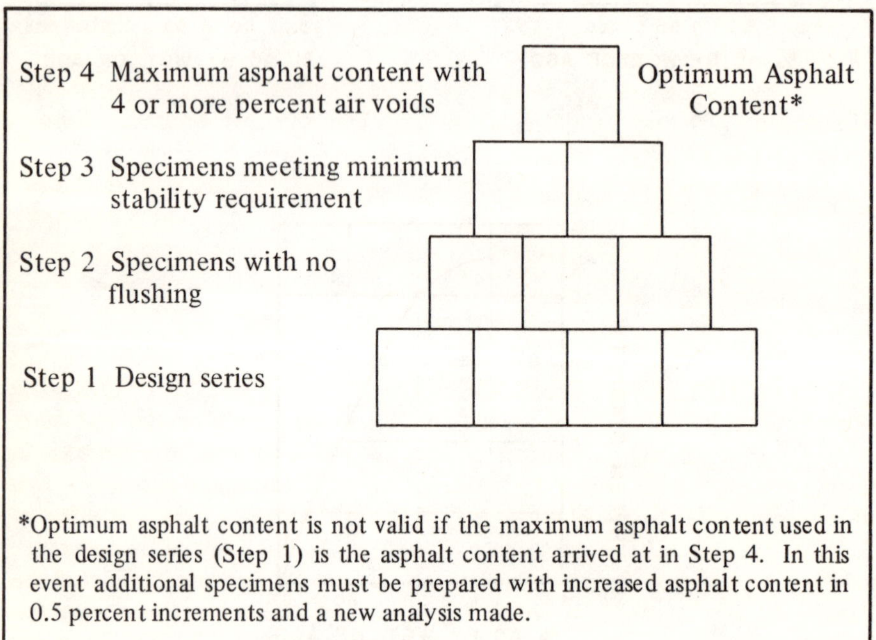

Figure IV-17–Procedures for selecting optimum asphalt content, Hveem method of design

EXAMPLE

The following example illustrates determination of the optimum asphalt content for a Heavy Traffic Category pavement using typical mix design test data shown in Figure IV-15.

Step 4	5.0			
Step 3	5.0	5.5		
Step 2	5.0	5.5	6.0	
Step 1	5.0	5.5	6.0	6.5

(a) Enter on Step (1) 5.0, 5.5, 6.0 and 6.5

(b) Specimen D exhibits "Moderate" surface flushing. Enter on Step (2) 5.0, 5.5 and 6.0

(c) Specimens A and B are the highest asphalt content specimens from Step (2) that provide stabilometer values of 37 or more. Enter on Step (3) 5.0 and 5.5

(d) Specimen A is the highest asphalt content specimen that has at least 4 percent air voids. Enter on Step (4) 5.0. The optimum asphalt content is 5.0 percent

F. *Calibration of Hveem Stabilometer Device*

4.21 CHECKING DISPLACEMENT IN STABILOMETER–The stabilometer reading is influenced by the displacement in the stabilometer, i.e., the lateral expansion of specimen during test. An increase in displacement yields lower P_h readings (greater apparent stability). For this reason, the displacement must be maintained at a constant value of 2.00 ± 0.05 turns (0.4 cubic inch (6.6 cm^3) by the procedure described herein. Variations in displacement are caused by entrained air and the failure of the diaphragm to fit tightly against the specimen after the initial pressure of 5 psi (34.5 kPa) has been applied. The displacement in the stabilometer is checked as follows:

(a) On models having a valve, check displacement pump by closing pump valve and turning handwheel until tight. If more than one-eighth turn is required to bring handwheel tight, the pump contains air which must be removed. (Late model stabilometers do not have an angle valve.)

(b) Place stabilometer on adjustable stage and insert a "dummy" test specimen of steel or brass, 4.0-inch (101.6 mm) in diameter x 5.5-inches (139.7 mm) high.

(c) Tighten clamp at base of stabilometer, place stabilometer on compression machine, and apply a load of 1,000 pounds (4.45 kN) to top of dummy specimen.

(d) Open angle valve, turn displacement pump handle to right and force fluid into the cell until the stabilometer test gauge registers a pressure of 5 psi (34.5 kPa).

(e) Set the displacement pump dial gauge to zero with adjustment screw, turn handle of displacement pump to right at the rate of two turns per second until pressure on stabilometer test gauge reads 100 psi (690 kPa).

(f) Record displacement as the number of revolutions of the pump handle required to increase the stabilometer test gauge from 5 psi (34.5 kPa) to 100 psi (690 kPa) [0.1 inch (2.5 mm) dial reading is equivalent to one turn displacement].

(g) The stabilometer test gauge should remain stationary at 100 psi (690 kPa). If it visibly retreats, there is a leak that must be located and sealed.

(h) Volume of displacement in cubic inches (cm^3) may be determined by multiplying the turns of displacement by 0.2 (3.3).

(i) If the displacement does not check within ± 0.05 turns of the required 2.00 [0.4 cubic inch (6.6 cm^3)] an adjustment is necessary; air is added to the system to increase the displacement, or removed from the system to decrease the displacement.

(j) To adjust displacement to the required value, turn pump handle to right until a pressure of 100 psi (690 kPa) is obtained on test gauge, then turn pump handle to left two turns. Either add or remove air until a pressure of 5 psi (34.5 kPa) is indicated on test gauge. Air may be added to air cell by means of the rubber bulb provided for this purpose. Continue check and adjustment until the required value of displacement is obtained.

4.22 FILLING STABILOMETER WITH LIQUID–If the displacement pump is of the rubber gasket "O" ring type, it is recommended that the stabilometer be filled with a machine oil having a Saybolt-Furol viscosity of approximately sixty seconds at 100°F (38°C). If the displacement pump is of the packing gland type, the stabilometer should be filled with a solution of 50 percent glycerine and 50 percent distilled water.

It is very important that all air, even the smallest bubbles, be excluded from the system. On models equipped with angle valve, the displacement pump should be filled after closing the angle valve connecting the inner chamber. Remove the small plug in the center of the displacement pump handle, and screw the handle out about 3 inches (76 mm). Fill with liquid and remove all air, then replace plug and loosen the packing gland nut. Screw in displacement pump until liquid oozes out around packing gland nut. Next, remove the small plug in the angle valve and, with valve closed, screw the pump in until all air is removed from angle valve. Remove the filling plug opposite the test gauge and, with the stabilometer resting on the face of the testing gauge, fill with liquid. The rubber diaphragm should be tapped lightly with fingers to remove any air. The machine should be rocked about to eliminate bubbles near the opening.

The above procedure should also be followed for stabilometers that have had the displacement pump converted from a packing gland to an "O" ring type and the oil described above used as the liquid medium.

Chapter V

ANALYSIS OF COMPACTED PAVING MIXTURES

5.01 GENERAL–The analytical procedures described in this chapter apply either to paving mixtures that have been compacted in the laboratory, or to undisturbed samples that have been cut from a pavement in the field. When a paving mixture is compacted in the laboratory, the compactive effort should provide a density equal to the density *the mixture will ultimately attain under traffic* following compaction by rolling during construction.

By analyzing a compacted paving mixture for air voids (V_a), voids in the mineral aggregate (VMA), and effective asphalt content (P_{be}) some indication of the probable service performance of the pavement is provided. The efficacy of compaction, either during construction or after years of service can be determined by comparing the specific gravity of an undisturbed sample cut from a pavement with the laboratory compacted specific gravity of the same paving mixture.

Three different aggregate specific gravity values are used at the present time for analyzing paving mixtures, bulk specific gravity (G_{sb}), apparent specific gravity (G_{sa}), and effective specific gravity (G_{se}).

The Asphalt Institute recommends that VMA values for compacted paving mixtures should be calculated in terms of the aggregate's bulk specific gravity, G_{sb}. The bulk specific gravity, with an allowance for the portion of the asphalt binder lost by absorption into the aggregate particles, or the equivalent of this procedure, should be the basis for calculating the air voids in a compacted asphalt paving mixture. The effective asphalt content (P_{be}) of a paving mixture is the total asphalt content minus the portion of the total asphalt content that is lost by absorption into the aggregate particles.

Values for voids in the mineral aggregate (VMA) and air voids (V_a) are expressed as percent by volume of the sample of paving mixture being analyzed. Depending on how asphalt content is specified, the effective asphalt content may be expressed either as percent by weight of the total weight of the sample of paving mixture being examined, or as percent by weight of the aggregate in the paving mixture sample.

Because air voids and voids in mineral aggregate are volume quantities and therefore cannot be weighed, a paving mixture must first be designed or analyzed on a volume basis. For design purposes, this volume approach can easily be changed over to a weight basis to provide a job-mix formula.

Table V-1 illustrates that the type of aggregate specific gravity employed for the analysis of a given sample of compacted paving mixture can have a very marked effect on the values reported for air voids and for voids in mineral aggregate. These differences are enough to make it appear that a mixture may satisfy the design criteria for air voids and VMA if one aggregate specific gravity is used for analysis, but that it fails to meet them if another specific gravity is employed. Consequently, specifications for paving mixture design and construction should be quite clear in stating precisely how air voids and VMA are to be determined.

5.02 DEFINITIONS—Mineral aggregate is porous and can absorb water and asphalt to a variable degree. Furthermore, the ratio of water to asphalt absorption varies with each aggregate. Three methods of measuring aggregate specific gravity that are in current use take these variations into consideration. They are ASTM bulk, ASTM apparent, and effective specific gravities, and these are defined as follows:

> *Bulk Specific Gravity,* G_{sb}—the ratio of the weight in air of a unit volume of a permeable material (including both permeable and impermeable voids normal to the material) at a stated temperature to the weight in air of equal density of an equal volume of gas-free distilled water at a stated temperature. *See* Figure V-1.
>
> *Apparent Specific Gravity,* G_{sa}—the ratio of the weight in air of a unit volume of an impermeable material at a stated temperature to the weight in air of equal density of an equal volume of gas-free distilled water at a stated temperature. *See* Figure V-1.
>
> *Effective Specific Gravity,* G_{se}—the ratio of the weight in air of a unit volume of a permeable material (excluding voids permeable to asphalt) at a stated temperature to the weight in air of equal density of an equal volume of gas-free distilled water at a stated temperature. *See* Figure V-1.

As the terms *effective asphalt content* (P_{be}), *air voids* (V_a), and *voids in the mineral aggregate* (VMA) will be used throughout this chapter, they should be defined.

> *Voids in the Mineral Aggregate,* VMA—the volume of intergranular void space between the aggregate particles of a compacted paving mixture that includes the air voids and the effective asphalt content, expressed as a percent of the total volume of the sample. *See* Figure V-1.

Effective Asphalt Content, P_{be}—the total asphalt content of a paving mixture minus the portion of asphalt that is lost by absorption into the aggregate particles. *See* Figure V-1.

Air Voids, V_a—the total volume of the small pockets of air between the coated aggregate particles throughout a compacted paving mixture, expressed as percent of the bulk volume of the compacted paving mixture. *See* Figure V-1.

TABLE V-1—INFLUENCE OF TYPE OF SPECIFIC GRAVITY ON DETERMINATION OF VMA AND AIR VOIDS

Bulk Specific Gravity of Compacted Mixture, G_{mb} 2.436
Density of Compacted Mixture, W_{pcf}, lb per cu ft (kg/m^3) 152.0 (2435)
Asphalt Content, percent by weight of total mix 5.93
Asphalt Absorbed by Aggregate Particles, W_{ba} lb per 100 lb (kg/45 kg) Aggregate .... 0.8 (.36)
Specific Gravity of Asphalt Cement, G_b 1.011

Laboratory Number	Specific Gravity Employed For Aggregate		Allowance For Absorption of Asphalt by Aggregate	Void Properties Compacted Mixture	
				Percent Voids in Mineral Aggregate	Percent Air Voids
No. 1	ASTM Bulk	2.651	Yes	13.56	1.08
No. 2	ASTM bulk	2.651	No	13.56	-0.73
No. 3	ASTM bulk (sat. surf. dry)	2.716	Yes	15.63	3.15
No. 4	ASTM bulk (sat. surf. dry)	2.716	No	15.63	1.34
No. 5	ASTM apparent	2.834	No	19.14	4.85
No. 6	Effective	2.708	No	15.37	1.08

Figure V-2 depicts the volume relationships between aggregate, air voids, voids in mineral aggregate, total asphalt content, asphalt lost by absorption into the aggregate particles, and effective asphalt content.

NOTE:

The accuracy of determinations of specific gravity for mix design is important.

Unless specific gravities are determined to four significant figures (three decimal places) an error in air voids value of as much as 0.8 percent can occur. Therefore, The Asphalt Institute recommends the use of weigh scales whose sensitivity will allow the weighing of a mix batch of from 1000 to 5000 grams to an accuracy of 0.1 gram.

Additionally desk calculators rather than slide rules should be used for all calculations.

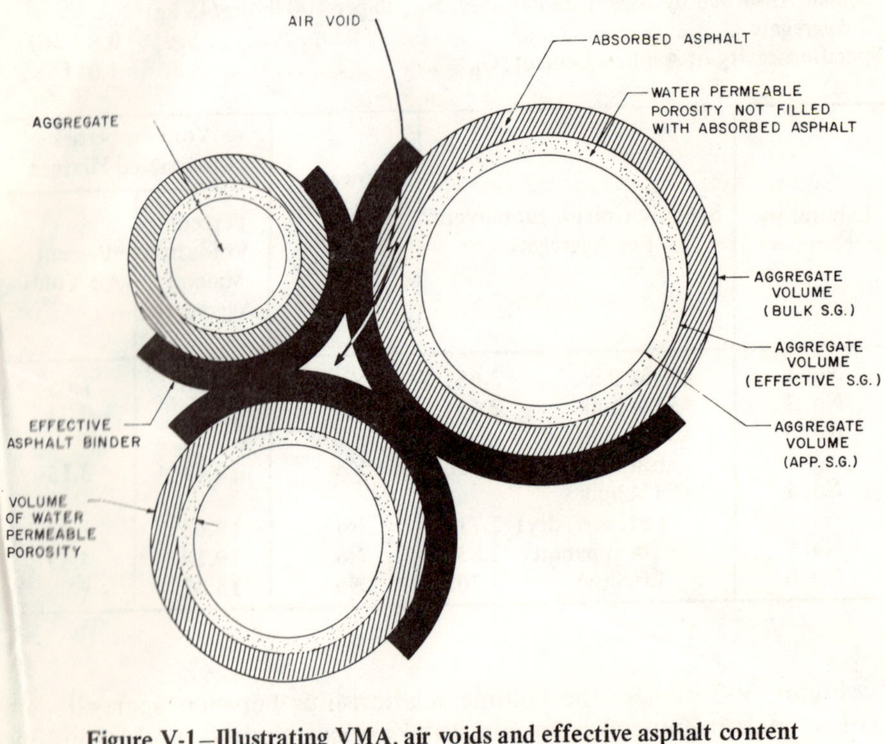

Figure V-1—Illustrating VMA, air voids and effective asphalt content in compacted asphalt paving mixture

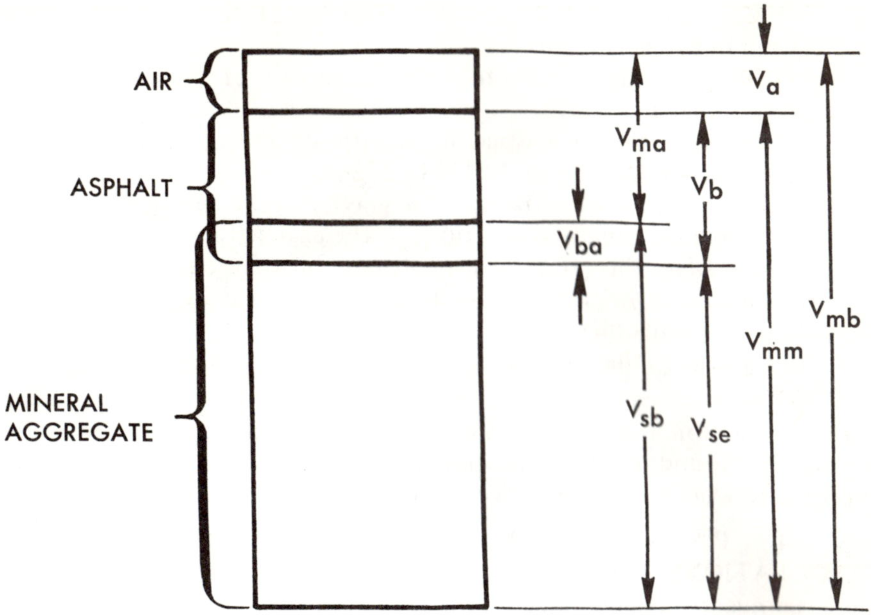

V_{ma} = Volume of voids in mineral aggregate
V_{mb} = Bulk volume of compacted mix
V_{mm} = Voidless volume of paving mix
V_a = Volume of air voids
V_b = Volume of asphalt
V_{ba} = Volume of absorbed asphalt
V_{sb} = Volume of mineral aggregate (by bulk specific gravity)
V_{se} = Volume of mineral aggregate (by effective specific gravity)

Figure V-2–Representation of volumes in a compacted asphalt specimen

5.03 OUTLINE OF PROCEDURE FOR ANALYZING A COMPACTED PAVING MIXTURE–

(a) Measure the bulk specific gravities of the coarse aggregate (AASHO T 85 or ASTM C 127) and of the fine aggregate (AASHO T 84 or ASTM C 128).

(b) Measure the specific gravity of the asphalt cement (AASHO T 228 or ASTM D 70) and of the mineral filler (AASHO T 100 or ASTM D 854).

(c) Calculate the bulk specific gravities of the aggregate combination in the paving mixture.

(d) Measure the maximum specific gravity of the loose paving mixture (ASTM D 2041).

(e) Measure the bulk specific gravity of the compacted paving mixture (ASTM D 1188 or ASTM D 2726).

(f) Calculate the effective specific gravity of the aggregate.

(g) Calculate asphalt absorptions of the aggregate.

(h) Calculate the effective asphalt content of the paving mixture.

(i) Calculate the percent voids in the mineral aggregate in the compacted paving mixture.

(j) Calculate the percent air voids in the compacted paving mixture.

Equations for the above calculations are found in Articles 5.05 through 5.11 and their application may be expedited by use of the appropriate worksheet, Figure V-3 or V-4.

5.04 PAVING MIXTURE DATA FOR SAMPLE CALCULATIONS—The following tabulation illustrates the basic data for a sample of paving mixture. These design data are used in the sample calculation employed throughout the remainder of the chapter.

TABLE V-2—BASIC DATA FOR SAMPLE OF PAVING MIXTURE

(a) Constituents

Material	Specific Gravity				Mix Composition	
		Bulk	AASHO Method	ASTM Method	Percent By Weight of Total Mix	Percent By Weight of Total Aggregate
Asphalt Cement	1.010(G_b)	---	T 228	D 70	6.96(P_b)	7.48(P_b)
Coarse Aggregate		2.606(G_1)	T 85	C 127	51.45(P_1)	55.30(P_1)
Fine Aggregate		2.711(G_2)	T 84	C 128	34.24(P_2)	36.80(P_2)
Mineral Filler	2.697(G_3)	---	T 100	D 854	7.35(P_3)	7.90(P_3)

(b) Paving Mixture

Bulk specific gravity of compacted paving mixture sample, G_{mb}

(ASTM D 2726) ——— 2.344

Maximum specific gravity of paving mixture sample, G_{mm}

(ASTM D 2041) ——— 2.438

5.05 BULK SPECIFIC GRAVITIES OF AGGREGATE–When the total aggregate consists of separate fractions of coarse aggregate, fine aggregate, and mineral filler, all having different specific gravities, the bulk and apparent specific gravities for the total aggregate are calculated as follows:

(a) *Bulk Specific Gravity*, G_{sb}

$$G_{sb} = \frac{P_1 + P_2 + \ldots + P_n}{\frac{P_1}{G_1} + \frac{P_2}{G_2} + \ldots + \frac{P_n}{G_n}} \tag{1}$$

where

G_{sb} = bulk specific gravity for the total aggregate

P_1, P_2, P_n = percentages by weight of aggregates, 1, 2, n; and

G_1, G_2, G_n = bulk specific gravities of aggregates 1, 2, n.

The bulk specific gravity of mineral filler is difficult to determine accurately at the present time. However, if the apparent specific gravity of the filler is used instead, the error is usually negligible.

Calculation using the data in Table V-2:

$$G_{sb} = \frac{51.450 + 34.240 + 7.350}{\frac{51.450}{2.606} + \frac{34.240}{2.711} + \frac{7.350}{2.697}} = \frac{93.040}{19.743 + 12.630 + 2.725} = 2.651$$

5.06 EFFECTIVE SPECIFIC GRAVITY OF AGGREGATE–When based on the maximum specific gravity of a paving mixture, G_{mm}, the effective specific gravity of G_{se}, of the aggregate includes all void spaces in the aggregate particles except those that absorb asphalt. It is determined as follows:

$$G_{se} = \frac{P_{mm} - P_b}{\frac{P_{mm}}{G_{mm}} - \frac{P_b}{G_b}} \tag{2}$$

where

G_{se} = effective specific gravity of aggregate

P_{mm} = total loose mixture, percent by total weight of mixture = 100 percent

P_b = asphalt, percent by total weight of mixture

G_{mm} = maximum specific gravity of paving mixture (no air voids), ASTM D 2041 (Refer to Article 5.07)

G_b = specific gravity of asphalt

Calculation using the data in Table V-2:

$$G_{se} = \frac{100 - 6.960}{\frac{100}{2.438} - \frac{6.960}{1.010}} = \frac{93.040}{34.126} = 2.726$$

NOTE:

The volume of asphalt binder absorbed by an aggregate is almost invariably less than the volume of water absorbed. Consequently, the value for the effective specific gravity of an aggregate should be between its bulk and apparent specific gravities. When the effective specific gravity falls outside these limits, its value must be assumed to be incorrect. The calculations, the maximum specific gravity of the total mix by ASTM D 2041, and the composition of the mix in terms of aggregate and total asphalt content should then be re-checked for the source of the error.

If the apparent specific gravity of the coarse aggregate is 2.759 and the apparent specific gravity of the fine aggregate is 2.905 for the sample of paving mix in Table V-2, the apparent specific gravity, G_{sa}, of the total aggregate can be calculated by the same formula as the bulk by using the apparent specific gravity of each aggregate constituent. For this example, then, the calculated apparent specific gravity, G_{sa}, is

$$G_{sa} = \frac{51.450 + 34.240 + 7.350}{\frac{51.450}{2.759} + \frac{34.240}{2.905} + \frac{7.350}{2.697}} = \frac{93.040}{33.159} = 2.806$$

In the example the three specific gravities are as follows:

Bulk Specific Gravity, G_{sb}	2.651
Effective Specific Gravity, G_{se}	2.726
Apparent Specific Gravity, G_{sa}	2.806

5.07 MAXIMUM SPECIFIC GRAVITIES OF MIXTURES WITH DIFFERENT ASPHALT CONTENTS—In designing a paving mixture with a given aggregate, the maximum specific gravities, G_{mm}, at different asphalt contents are needed to calculate the percentage of air voids for each asphalt content. While the same maximum specific gravity can be determined for each asphalt content by ASTM test method D 2041, the precision of the test is best when the mixture has close to the optimum asphalt content. Also, it is preferable to measure the maximum specific gravity in duplicate or triplicate.

After averaging the results from these tests and calculating the effective specific gravity of the aggregate, the maximum specific gravity for any other asphalt content can be obtained as shown below. For all practical purposes, the effective specific gravity of the aggregate is constant because the asphalt absorption does not vary appreciably with variations in asphalt content.

$$G_{mm} = \frac{P_{mm}}{\dfrac{P_s}{G_{se}} + \dfrac{P_b}{G_b}} \qquad (3)$$

where

G_{mm} = maximum specific gravity of paving mixture (no air voids)

P_{mm} = total loose mixture, percent by total weight of mixture = 100 percent

P_s = aggregate, percent by total weight of mixture

P_b = asphalt, percent by total weight of mixture

G_{se} = effective specific gravity of aggregate

G_b = specific gravity of asphalt

Calculation using specific gravity data from Table V-2, effective specific gravity, G_{se}, determined in Article 5.06 and an asphalt content, P_b, of 7.46 percent:

$$G_{mm} = \frac{100}{\dfrac{92.540}{2.726} + \dfrac{7.460}{1.010}} = \frac{100}{41.333} = 2.419$$

5.08 ASPHALT ABSORPTION—Absorption is expressed as a percentage by weight of aggregate rather than as a percentage by total weight of mixture. Asphalt, P_{ba}, absorption is determined as follows:

Asphalt absorption –

$$P_{ba} = 100 \frac{G_{se} - G_{sb}}{G_{sb} G_{se}} G_b \qquad (4)$$

where

P_{ba} = absorbed asphalt, percent by weight of aggregate

G_{se} = effective specific gravity of aggregate

G_{sb} = bulk specific gravity of aggregate

G_b = specific gravity of asphalt

Calculation using bulk and effective specific gravities determined in Articles 5.05 and 5.06 and asphalt specific gravity from Table V-2.

$$P_{ba} = 100 \frac{2.726 - 2.651}{2.651 \times 2.726} 1.010 = 100 \frac{0.075}{7.227} 1.010 = 1.05$$

5.09 EFFECTIVE ASPHALT CONTENT OF A PAVING MIXTURE–The effective asphalt content, P_{be}, of a paving mixture is the total asphalt content minus the quantity of asphalt lost by absorption into the aggregate particles. It is the portion of the total asphalt content that remains as a coating on the outside of the aggregate particles, and is the asphalt content on which service performance of an asphalt paving mixture depends. The formula is:

$$P_{be} = P_b - \frac{P_{ba}}{100} P_s \qquad (5)$$

where

P_{be} = effective asphalt content, percent by total weight of mixture

P_b = asphalt, percent by total weight of mixture

P_{ba} = absorbed asphalt, percent by weight of aggregate

P_s = aggregate, percent by total weight of mixture

Calculation using data from Table V-2 and Article 5.08:

$$P_{be} = 6.96 - \frac{1.05}{100} 93.04 = 6.96 - 0.98 = 5.98$$

5.10 PERCENT VMA IN COMPACTED PAVING MIXTURE—The voids in the mineral aggregate, VMA, are defined as the intergranular void space between the aggregate particles in a compacted paving mixture that includes the air voids and the effective asphalt content, expressed as a percent of the total volume. The VMA is calculated on the basis of the bulk specific gravity of the aggregate and is expressed as a percentage of the bulk volume of the compacted paving mixture. Therefore, the VMA can be calculated by subtracting the volume of the aggregate determined by its bulk specific gravity from the bulk volume of the compacted paving mixture. The method of calculation is illustrated as follows:

If mix composition is determined as percent by weight of total mixture:

$$VMA = 100 - \frac{G_{mb} P_s}{G_{sb}} \qquad (6)$$

where

VMA = voids in mineral aggregate (percent of bulk volume)

G_{sb} = bulk specific gravity of aggregate

G_{mb} = bulk specific gravity of compacted mixture (ASTM D 2726)

P_s = aggregate, percent by total weight of mixture

Calculation using data from Table V-2 and Article 5.05:

$$VMA = 100 - \frac{2.344 \times 93.04}{2.651} = 100 - 82.27 = 17.73$$

Or if mix composition is determined as percent by weight of aggregate:

$$VMA = 100 - \frac{G_{mb}}{G_{sb}} \times \frac{100}{100 + P_b} 100 \qquad (7)$$

where

P_b = asphalt, percent by weight of aggregate

Calculation using data from Table V-2 and Article 5.05:

$$VMA = 100 - \frac{2.344}{2.651} \times \frac{100}{100 + 7.48} 100 = 100 - 82.27 = 17.73$$

5.11 CALCULATION OF PERCENT AIR VOIDS IN COMPACTED MIXTURE—The air voids, P_a, in a compacted paving mixture consist of the small air spaces *between* the coated aggregate particles. The percentage of air voids in a compacted mixture can be determined by the following equation:

$$P_a = 100 \frac{G_{mm} - G_{mb}}{G_{mm}} \qquad (8)$$

where

P_a = air voids in compacted mixture, percent of total volume

G_{mm} = maximum specific gravity of paving mixture (as determined in Article 5.07 or as determined directly for a paving mixture by ASTM Method D 2041)

G_{mb} = bulk specific gravity of compacted mixture

Calculation using data from Table V-2:

$$P_a = 100 \frac{2.438 - 2.344}{2.438} = \frac{9.400}{2.438} = 3.85$$

WORKSHEET FOR ANALYSIS OF COMPACTED PAVING MIXTURE

(Analysis by weight of total mixture)

Sample: *FI-005-5 (4)* Date: *October 10, 1973*

Identification: *REX-EAST*

Composition of Paving Mixture

Constituent Material		Specific Gravity, G	Bulk		Mix Composition, % by weight of Total Mix., P				
					Mix or Trial Number				
					1	2	3	4	5
1. Coarse Aggregate	G_1	2.759	2.606	P_1			51.45		
2. Fine Aggregate	G_2	2.905	2.711	P_2			34.24		
3. Mineral Filler	G_3	2.697	--	P_3			7.35		
4. Total Aggregate	G_s	--	--	P_s			93.04		
5. Asphalt Cement	G_b	1.010	--	P_b			6.96		

	Equation*	1	2	3	4	5
6. Bulk Sp. Gr. (G_{sb}), total aggregate	(1)			2.651		
7. Max. Sp. Gr. (G_{mm}), paving mix (ASTM D2041)	(3)			2.438		
8. Bulk Sp. Gr. (G_{mb}), compacted mix (ASTM D2726)	–			2.344**		
9. Effective Sp. Gr. (G_{se}), total aggregate	(2)			2.726		
10. Absorbed Asphalt (P_{ba}), % by wgt. total agg.	(4)			1.05		
CALCULATIONS						
11. Effective Asphalt Content (P_{be}) = Line 5 P_b − (Line 10 x 0.01 x Line 4 P_s)	(5)			5.98		
12. VMA = $100 - \frac{\text{Line 8} \times \text{Line 4 } P_s}{\text{Line 6}}$	(6)			17.73		
13. Air Voids (Pa) = $100 \left(\frac{\text{Line 7} - \text{Line 8}}{\text{Line 7}}\right)$	(8)			3.85		
14. Stability (Marshall Hveem)						
15. Flow (Marshall)						
16. Swell (Hveem)						

*Equations from Chapter V, MS-2

**Average of three or more samples for Marshall Method or one sample for Hveem Method

Figure V-3—Worksheet: Analysis by weight of total mixture

WORKSHEET FOR ANALYSIS OF COMPACTED PAVING MIXTURE

(Analysis by weight of aggregate)

Sample: FI-005-5(4) Date: October 10, 1973

Identification: REX - EAST

Composition of Paving Mixture

Constituent Material		Specific Gravity, G	Bulk	Equation*		Mix 1	Mix 2	Mix 3	Mix 4	Mix 5
1. Coarse Aggregate	G_1	2.759	2.606		P_1			55.30		
2. Fine Aggregate	G_2	2.905	2.711		P_2			36.80		
3. Mineral Filler	G_3	2.697	--		P_3			7.90		
4. Total Aggregate	G_s	--	--		P_s			100.00		
5. Asphalt Cement	G_b	1.010	--		P_b			7.48		
6. Bulk Sp. Gr. (G_{sb}), total aggregate				(1)				2.651		
7. Max. Sp. Gr. (G_{mm}), paving mix (ASTM D2041)				(3)				2.438		
8. Bulk Sp. Gr. (G_{mb}), compacted mix (ASTM D2726)				–				2.399**		
9. Effective Sp. Gr. (G_{se}), total aggregate				(2)				2.726		
10. Absorbed Asphalt (P_{ba}), % by wgt. total agg.				(4)				1.05		
CALCULATIONS										
11. Effective Asphalt Content (P_{be}) = Line 5 P_b – Line 10				(5)				6.43		
12. VMA = $100 - \frac{\text{Line 8}}{\text{Line 6}} \times \frac{100}{100 + \text{Line 5 } P_b} 100$				(7)				17.73		
13. Air Voids (P_a) = $100 \left(\frac{\text{Line 7} - \text{Line 8}}{\text{Line 7}} \right)$				(8)				3.85		
14. Stability (Marshall Hveem)										
15. Flow (Marshall)										
16. Swell (Hveem)										

Mix Composition, % by weight of Aggregate, P (Mix or Trial Number 1–5)

*Equations from Chapter V, MS-2

**Average of three or more samples for Marshall Method or one sample for Hveem Method

Figure V-4–Worksheet: Analysis by weight of aggregate

Appendix

GRADATION ANALYSIS OF AGGREGATES

A.01 GENERAL–The analysis of aggregate gradations and the combining of aggregates to obtain the desired gradation are important steps in hot-mix design. The aggregate gradation must meet the gradation requirements of the project specifications and yield a mix design that meets the criteria of the design method. And, the gradation should be made up of the most economical proper aggregates to be found.

This appendix outlines a recommended schedule for analyzing aggregates for asphalt paving mix design. The methods illustrated by these examples are applicable to blending and adjusting the aggregate gradation in laboratory control of the mix, in production control of aggregates, and in plant control during construction.

A.02 SCHEME FOR ANALYZING AGGREGATE FOR PRELIMINARY MIX DESIGNS–For preliminary mix designs, the scheme for analyzing aggregate will be governed, to some extent, by the method of producing the gradation during construction. One of several methods may be used, depending on the type of local aggregate sources and the requirements of the project specifications. Aggregates from local roadside pit sources are frequently produced either as a "single aggregate," or as a "major aggregate" supplemented by minor additions of filler or "sized aggregates." The roadside production of aggregates often will require a certain amount of "waste" to obtain the desired grading. On the other hand, aggregates from commercial producers are usually supplied as "sized" aggregates, in which case all of the various sizes required are blended, with or without filler, to produce the aggregate gradation desired.

The following outlines the scheme of laboratory operations that normally applies to the testing of aggregates for preliminary or exploratory mix designs. It will be necessary at times to modify this plan to meet the testing requirements that arise during the progress of the mix design.

(a) Dry all aggregate samples to constant weight at 105°C to 110°C. Separate pans should be used for each aggregate sample.

(b) Perform washed sieve analysis and specific gravity tests on representative samples including filler from each proposed aggregate source.

(c) Compute blend of aggregates required to produce the desired mix gradation, using the gradation for each aggregate source (adjusted for waste if required). For preliminary or exploratory mix designs it is usually advisable to make the initial trial mixes using an aggregate gradation that approaches the median of the specification limits.

(d) Blend the aggregates from each source in proportions and amounts to produce the full amount of aggregate required for one trial mix series of the desired gradation. It is usually not advisable to blend in the filler material at this time. For example, the gradation analysis of Aggregates A and B and Fine Sand C indicates that these three aggregates should be blended 50 percent, 40 percent, and 10 percent, respectively, to produce the required gradation. Thus, if a total of 20 kg of aggregate is required for one trial mix series, 10 kg of Aggregate A and 8 kg of Aggregate B should be blended together as a uniform mixture. The remaining 2 kg of Fine Sand C, however, should be blended directly into the batch mixes.

(e) Separate the blended aggregates by dry sieving into fractions. The following sizes are recommended as generally adequate:

plus 3/4 in. (19.0 mm)
3/4 in. (19.0 mm) to 3/8 in. (9.5 mm)
3/8 in. (9.5 mm) to No. 4
No. 4 to No. 8
minus No. 8

The size separations may vary depending on the grading of the aggregate and on the aggregate heating and screening facilities available.

(f) Perform a washed sieve analysis using a representative sample from each sized fraction of aggregate and a representative sample filler.

(g) Compute the blend proportions and batch weights of the sized aggregates and filler required to produce one batch mix of the desired gradation. As a matter of practical convenience it is preferable to use the same weight of aggregate for each batch in a trial mix series.

(h) Prepare test specimens and design mix in accordance with the procedure prescribed for the particular mix design method being used.

A.03 PLAN FOR AGGREGATE ANALYSIS FOR JOB MIX FORMULAS—For determining job mix formulas, the plan for aggregate analysis will be governed to some extent by the number of aggregate stockpiles and the number of bin separations being used in the production of the paving mix. The purpose of this phase of mix design is to establish the job mix formula that defines the actual gradation and asphalt content to be obtained in the finished construction.

The following outlines the laboratory operations that normally apply to the testing of aggregates for establishing job mix formulas. For the actual details of field sampling, plant calibration and control, refer to The Asphalt Institute's *Asphalt Plant Manual* (MS-3). It will be necessary at times to modify this plan to suit the control features of the central proportioning and mixing plant.

(a) Secure representative samples from each aggregate stockpile including filler, to be used in the production of the asphalt paving mix.

(b) Dry all aggregate samples to constant weight at 105°C to 110°C. Separate pans should be used for each aggregate sample.

(c) Perform washed sieve analysis and specific gravity tests on each representative sample from the respective stockpiles.

(d) Compute the blend of aggregates required to produce the desired mix gradation, using the full gradation for each separate aggregate.

(e) Adjust the cold aggregate feeder controls to obtain the desired aggregate blend and combined gradation.

(f) Perform washed sieve analysis on a representative sample of each aggregate size separation produced in the plant. These samples should be lifted from the plant bins only after the gradation unit of the plant has attained normal operation.

(g) Compute the blend proportions and batch weights of the sized aggregates and filler required to produce one batch mix (for laboratory design tests) of the desired gradation. It is preferable to use the same weight of aggregate for each batch in a trial mix series.

(h) Prepare test specimens and design mix in accordance with the procedure prescribed for the mix design method being used.

(i) Adjust plant controls to obtain the asphalt content and blend proportions of aggregates desired in the paving mix.

(j) Perform washed sieve analysis on representative samples of extracted aggregate as required to determine that the desired gradation is being obtained in the finished mix.

A.04 BLENDING AGGREGATES BY WEIGHT—Determining the proportions of two or more aggregates, to blend for a gradation within the specification limits, is largely a matter of trial and error, although graphical methods may often be used to advantage. It is always desirable first to plot the sieve analyses for all aggregates to be used as shown in Figures A-1 and A-2. In this way it is often possible to make a visual estimate of the blend proportions required.

A.05 BASIC FORMULA–Regardless of the number of aggregates combined or of the method by which the proportions are determined, the basic formula expressing the combination is

$$P = Aa + Bb + Cc \text{ etc.} \tag{1}$$

where

P = the percentage of material passing a given sieve for the combined aggregates A,B,C, etc.;

A,B,C, etc. = percentage of material passing a given sieve for aggregates A,B,C, etc.; and

a,b,c, etc. = proportions of aggregates, A,B,C, etc. used in the combination and where the total = 1.00.

The combined percentages, P, in Eq. 1 should closely agree with the desired percentages for the combination for the different sieve sizes. None should fall outside the established grading specification limits. Obviously, there may be several acceptable combinations. An optimum combination would be one in which the percentages of the blend are in as close agreement as possible to the desired percentages originally set up.

Sophisticated mathematical procedures have been developed that will determine an optimum combination of aggregates. Even with a calculator, however, these methods are involved and time-consuming. Usually a trial-and-error method guided by a certain amount of reasoning is the easiest procedure to determine a satisfactory combination.

A.06 GRAPHICAL SOLUTIONS–Graphical methods have been devised for determining combinations of aggregates to obtain a desired gradation. Like mathematical methods, some graphical methods are quite complicated. As the number of aggregates to be combined is increased, the graphical method becomes even more complicated. For two and sometimes three aggregate materials, graphical solutions may be used to advantage over trial-and-error methods. In other instances, graphical methods may be used to indicate the starting point for trial-and-error solutions.

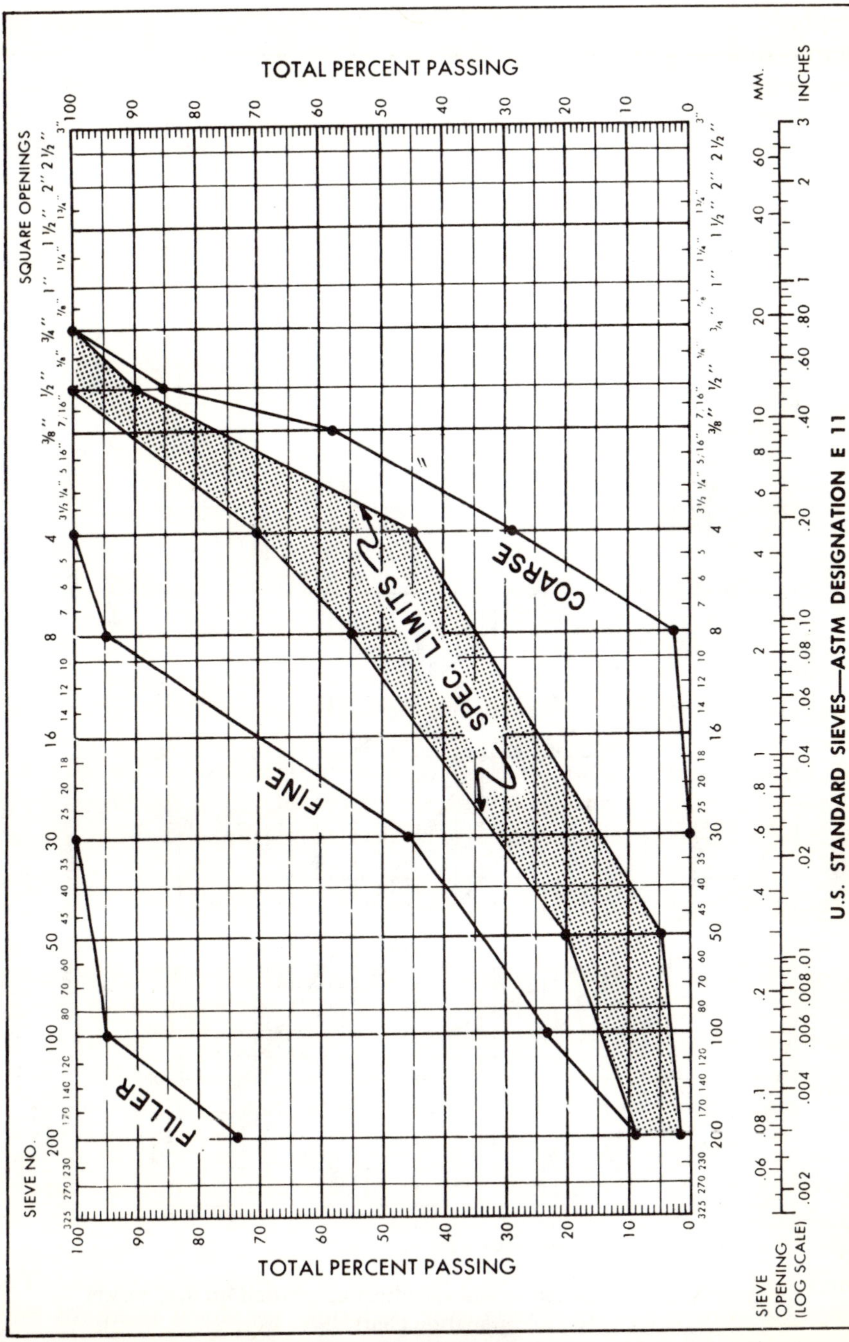

Figure A-1–Job aggregates and specification plotted on conventional aggregate grading chart

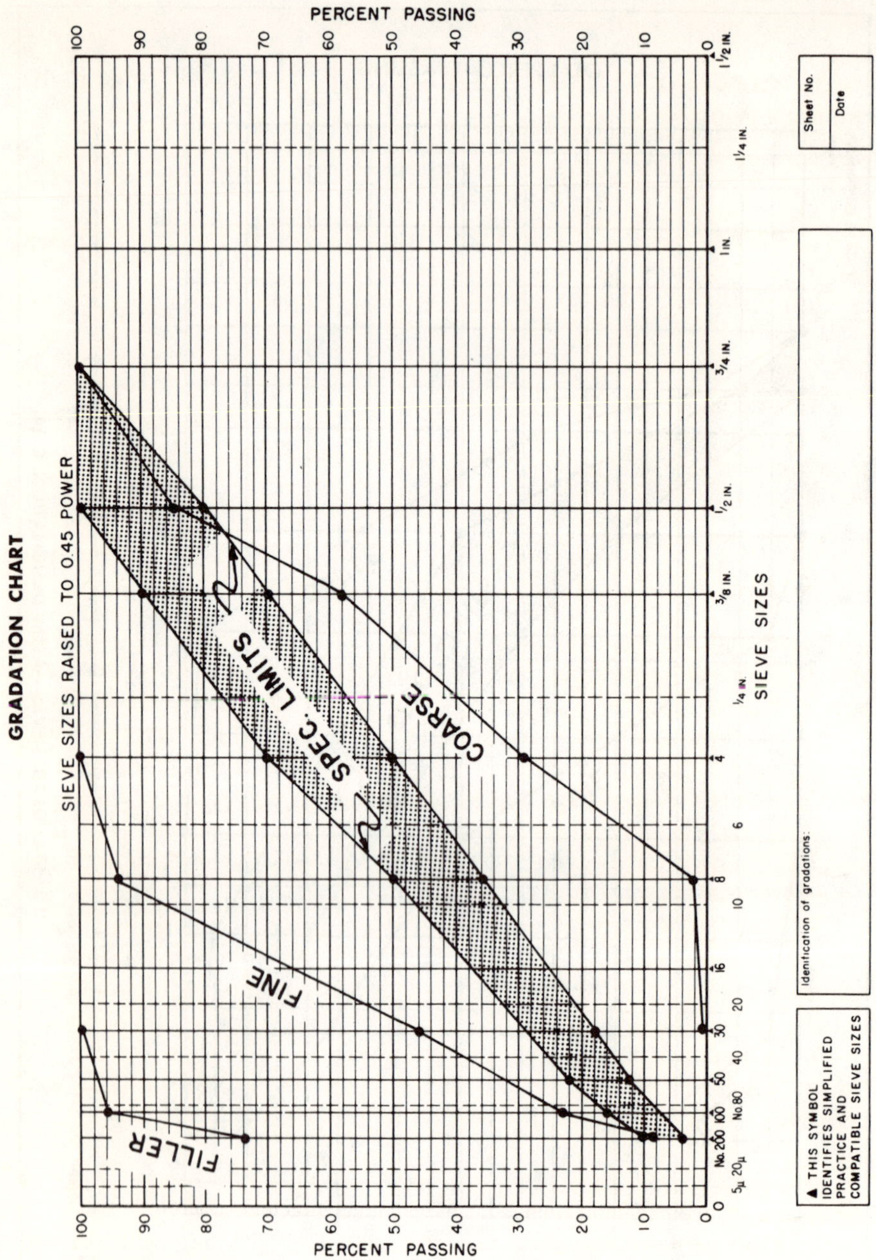

Figure A-2–Job aggregates and specification plotted on 0.45 power gradation chart

A.07 COMBINING TWO AGGREGATES–The basic formula for combining two aggregates is:

$$P = Aa + Bb \qquad (2)$$

Since a + b = 1, then a = 1 - b. Substituting this into Eq. 2 and solving for b:

$$b = \frac{P - A}{B - A} \qquad (3)$$

An expression for a, can also be found:

$$a = \frac{P - B}{A - B} \qquad (4)$$

Assume that a single aggregate stockpile is to be blended with sand to meet grading requirements for an asphalt paving mixture. These are given in Figure A-3a as aggregates A and B. To make a determination:

1. Examine the two gradations to determine which aggregate will contribute most of the material for certain sizes. In this case, most of the minus No. 8 aggregate will be furnished by aggregate B.
2. Using the percentages for the No. 8 sieve and substituting into Eq. 3, the proportions are determined to meet the midpoint of the specification (Figure A-3b).
3. Inspection of the blended gradation shows the percent passing No. 200 close to the lower specification limit. Increase the proportion of aggregate B (in this case to 0.55) and compute the gradation of the revised blend (Figure A-3c).
4. Inspection now shows the gradation is critical on the No. 30 sieve. Reduce the proportion of aggregate B to 0.52 or 0.53 and compute the gradation of the revised blend (Figure A-3d).

The two aggregates may be combined graphically (Figure A-4) as follows:

1. The percents passing the various sizes for aggregate A are plotted on the right-hand vertical scale (representing 100 percent aggregate A).
2. The percents passing the various sizes for aggregate B are plotted on the left-hand vertical scale (representing 100 percent aggregate B).

PER CENT PASSING

Sieve	3/4"	1/2"	3/8"	No.4	8	30	50	100	200
Spec.	100	80-100	70-90	50-70	35-50	18-29	13-23	8-16	4-10
Aggr. A	100	90	59	16	3.2	1.1	0	0	0
Aggr. B	100	100	100	96	82	51	36	21	9.2

(a) Grading Specification and sieve analyses of aggregates

For (No.8), $b = \frac{P-A}{B-A} = \frac{42.5-3.2}{82-3.2} = 0.50, \quad a = 1-0.50 = 0.50$

Sieve	3/4"	1/2"	3/8"	No.4	8	30	50	100	200
.50 × A	50	45.0	29.5	8.0	1.6	0.6			
.50 × B	50	50.0	50.0	48.0	41.0	25.0	18.0	10.5	4.6
Total	100	95.0	79.5	56.0	42.6	25.6	18.0	10.5	4.6
Spec.	100	80-100	70-90	50-70	35-50	18-29	13-23	8-16	4-10

Minus No. 200 low, increase b to 0.55, a to 0.45

(b) First trial combination

Sieve	3/4"	1/2"	3/8"	No.4	8	30	50	100	200
.45 × A	45	40.5	26.6	7.2	1.4	0.5			
.55 × B	55	55.0	55.0	52.8	45.1	28.0	19.8	11.5	5.1
Total	100	95.5	81.6	60.0	46.5	28.5	19.8	11.5	5.1
Spec.	100	80-100	70-90	50-70	35-50	18-29	13-23	8-16	4-10

Minus No. 30 high, let b = 0.52, a = 0.48

(c) Second trial combination

Sieve	3/4"	1/2"	3/8"	No. 4	8	30	50	100	200
.48 × A	48	43.2	28.3	7.7	1.5	0.5	0	0	0
.52 × B	52	52	52	49.9	42.6	26.5	18.7	10.9	4.8
Total	100	95.2	80.3	57.6	44.1	27.0	18.7	10.9	4.8
Spec.	100	80-100	70-90	50-70	35-50	18-29	13-23	8-16	4-10

(d) Third trial combination

Figure A-3–Trial-and-error calculations for combining two aggregates

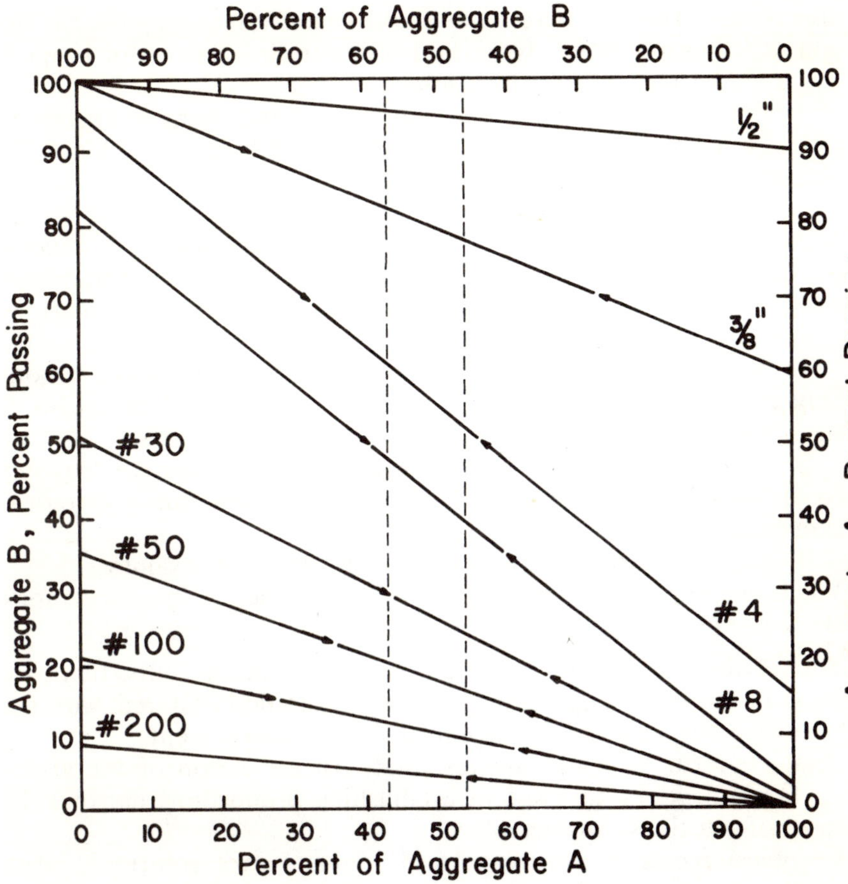

Figure A-4–Solution for proportioning two aggregates

3. Connect the points common to the same size with straight lines, and label.

4. For a particular size, indicate on the straight line where the line crosses the specification limits measured on the vertical scale. (Note that for the 3/8-in. (9.5 mm) size, two points are plotted on the line at 70 and 90 percent on the vertical scale.)

5. That portion of the line between the two points represents the proportions of aggregates A and B, measured on the horizontal scale, that will not exceed specification limits for that particular size.

6. The portion of the horizontal scale designated by two vertical lines, when projected vertically, is within specification limits for all sizes and represents the limits of proportions possible for satisfactory blends. In this case, 43 to 54 percent of aggregate A and 46 to 57 percent of aggregate B will meet specifications when blended. It can

also be seen that the percent of blended material passing the No. 30 and 200 sieves will be the critical or controlling values for keeping the blend within specification limits.

7. For blending, usually the midpoint of that horizontal scale is selected for the blend. In this case, 48 percent aggregate A and 52 percent aggregate B.

A.08 COMBINING THREE AGGREGATES—Assume that mineral filler, C, is to be blended with aggregates A and B to obtain a gradation meeting specification requirements. The specification and gradations are given in Figure A-5 where:

1. An inspection of the gradations indicates that there is a reasonably clean separation between the plus No. 8 sizes and minus No. 8 sizes. Aggregate A will furnish most of the plus No. 8 sizes.

2. Determine the approximate proportion of aggregate A required to obtain 42.5 percent passing the No. 8 sieve (midpoint of specification range), using Eq. 4. (*See* Figure A-5).

3. The percentages passing No. 200 sieve are examined next. Values are substituted into Eq. 1. The remainder of the calculations are self-explanatory.

Should the blended gradation exceed specification limits, that proportion in the blend apparently responsible should be altered, with the other proportions altered to make up a total of 100 percent.

Trial-and-error solutions are just that. An inspection of the gradations for indications to assist in establishing proportions narrows the number of solutions considerably.

Graphical methods usually help in trial-and-error solutions; when they do not, it is necessary to determine the proportions in cases of aggregate gradations having overlapping grading curves. Several graphical methods are possible, but this appears to be one of the more practical procedures. Each of the aggregates is divided into the following gradings:

(a) percent material retained on No. 8 sieve
(b) percent material passing No. 8 but retained on No. 200 sieve
(c) percent material passing No. 200 sieve.

The specification limits are divided in a somewhat similar manner: allowable percentage limits retained on a No. 8 sieve, and allowable percentage limits passing No. 200 sieve.

Points representing each of the three aggregate gradings are plotted on a chart, Figure A-6. Although only the portion retained on the No. 8 sieve and the portion passing the No. 200 sieve are used for each of the

PER CENT PASSING

Sieve	3/4"	1/2"	3/8"	No.4	8	30	50	100	200
Spec.	100	80-100	70-90	50-70	35-50	18-29	13-23	8-16	4-10
Aggr. A	100	90	59	16	3.2	1.1			
Aggr. B	100	100	100	96	82	51	36	21	9.2
Aggr. C	100	100	100	100	100	100	98	93	82

(a) Grading specification and gradations

(No. 8), $a = \frac{P-B}{A-B} = \frac{42.5-82}{3.2-82} = 0.50$

(No. 200), $P = Aa + Bb + Cc$

$7 = 0(0.50) + 9.2b + 82c, \quad b + c = 1 - 0.50 = 0.50$

$7 = 9.2(0.50 - c) + 82c$

$c = \frac{7-4.6}{82-9.2} = 0.03, \quad b = 0.50 - 0.03 = 0.47$

Sieve	3/4"	1/2"	3/8"	No.4	8	30	50	100	200
.50 × A	50	45.0	29.5	8.0	1.6	0.6			
.47 × B	47	47.0	47.0	45.1	38.5	24.0	16.9	9.9	4.3
.03 × C	3	3.0	3.0	3.0	3.0	3.0	3.0	2.8	2.5
Total	100	95.0	79.5	56.1	43.1	27.6	19.9	12.7	6-8
Spec.	100	80-100	70-90	50-70	35-50	18-29	13-23	8-16	4-10

(b) First trial solution for combining three aggregates

Note: To convert inches to millimeters multiply by 25.4

Figure A-5—Trial-and-error solution for combining three aggregates

three aggregates to locate points on the chart, the point designated A represents the coarse aggregate grading, B, the fine aggregate grading, and C, the mineral filler grading. Point S represents the middle of the specification grading band for material *retained* on the No. 8 sieve and *passing* the No. 200 sieve. Lines are drawn between Points A and S and between Points B and C. Lines AS is extended to Line BC to establish Point B′. Each line segment is determined by using the differences in percentages between terminal points. Then the percentage of each aggregate material needed for the blend is calculated using the following formulas derived from the basic formula (eq. 1).

$$a = \frac{\text{Line segment SB}'}{\text{Line AB}'}$$

$$b + c = 1.00 - a$$

$$c = (1.00 - a)\frac{\text{Line Segment } B'B}{\text{Line } CB}$$

$$b = 1.00 - (a + c)$$

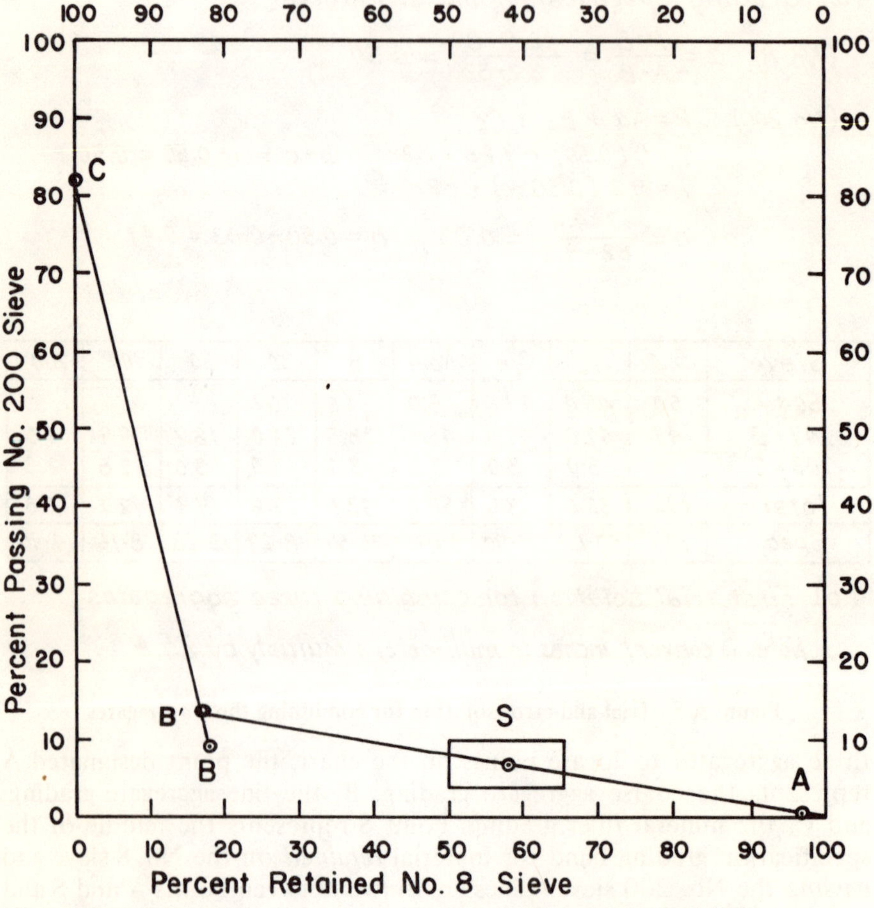

Figure A-6–Chart for estimating three aggregate blends

A.09 ADJUSTING FOR DIFFERENT SPECIFIC GRAVITIES–Aggregate gradations and grading curves are determined and expressed in percentages of total weight. Grading specifications, however, are established to meet volumetric requirements in the asphalt paving mix.

As long as the specific gravities of the combined aggregate materials are reasonably alike, the percentages by weight may be interpreted as percentages by volume for all practical purposes. When the specific gravities of the individual aggregates differ significantly (usually by 0.20 or more), the aggregate proportions should be adjusted. The adjustment is based on the fact that:

VOLUME X SPECIFIC GRAVITY = WEIGHT

Assume the following combination was calculated for three aggregates having different specific gravities:

Aggregate	Sp Gr	Proportion
A	1.00	0.52
B	2.00	0.45
C	3.00	0.03

Calculations for adjustments are shown in Figure A-7

Aggr.	Percent Vol.	Sp Gr	(1) Weight	(2) Percent Wt.
A	52.0	1.00	52.00	34.4
B	45.0	2.00	90.00	59.6
C	3.0	3.00	9.00	6.0
Total	100.0	---	151.00	100.0

(1) Weight = Vol. x Sp Gr

(2) Percent Wt. $= \dfrac{\text{Individual Weight, W}}{\text{Total Weight}} \times 100 = \dfrac{W}{151} \times 100$

Figure A-7—Adjusting percentages by volume to percentages by weight

A.10 ADJUSTING BY WASTING—Where the main source of aggregate is a local roadside pit, it is often the case that crusher-run aggregates are either coarser or finer than desired. If the gradation is coarser than desired, finer material can usually be blended with the crusher-run aggregate. But for gradations that have an excess of fines, the most economical adjustment is usually made by wasting a portion of the fine fraction. Most crushing plants will make the separations on the No. 4 (or possibly No. 8) screen. Where an excess on a smaller size occurs, the correction is made by wasting a portion of the minus No. 4 (or No. 8) fraction. The amount of waste is expressed as a percent, considering the total crusher-run material as 100 percent.

The formulas for analysis of gradations before and after wasting are as follows:

(Sizes above waste screen)

$$R_b = \frac{R_2 R_a}{R_1} \tag{5}$$

(Sizes below waste screen)

$$P_b = \frac{P_2 P_a}{P_1} \tag{6}$$

where

P_a, R_a = percent passing, or retained, of given size before wasting;
P_b, R_b = adjusted percent passing, or retained, of given size after wasting;
P_1, R_1 = percent passing, or retained, of waste size before wasting; and
P_2, R_2 = percent passing, or retained, of waste size after wasting.

The percent of waste, W, is found as follows:

$$W = \frac{(P_1 - P_2)100}{(100 - P_2)} \tag{7}$$

Assume that a single aggregate stockpile is being produced from a local roadside pit. The specification limits and crusher-run gradation are shown in Figure A-8a. Note that the No. 4 size is above specification limits and that the other percentages approach the upper limits of the specification. A portion of the minus No. 4 fraction will be wasted to reduce the percent passing No. 4 from 75 to 70. The adjusted percentages of the passing No. 4 sizes are found using Eq. 6 as shown. The percentages of the retained No. 4 sizes are first converted to percent-retained and the adjusted percent retained on No. 4 determined. The percentages of coarse sizes are found using Eq. 5 and reconverting to percent passing. The percent of waste of the passing No. 4 fraction is then found using Eq. 7.

Assume, in this case, that the No. 30 size exceeds specification limits. Therefore, a sufficient amount of the minus No. 4 fraction will be wasted to reduce the No. 30 from 31 percent to 28 percent (Figure A-8b). In this case, P_1 and P_2 for the fine fraction are values for the No. 30 sieve. When the adjusted percentages are obtained, R_1 and R_2 for the No. 4 sieve are used for the coarse fraction.

Sieve	3/4"	1/2"	3/8"	#4	8	30	100	200
Spec.	100	80-100	70-90	55-73	40-55	20-30	10-18	4-10
% Pass, Pa	100	98	87	75	54	28	17	9
% Ret, Ra	0	2	13	25				
Adj % Ret, Rb	0	2	16	30				
Adj % Pass, Pb	100	98	84	70	50	26	16	8.4

$$Pb = \frac{P_2}{P_1} Pa = \frac{70}{75} Pa = 0.93 Pa \qquad Rb = \frac{R_2}{R_1} Ra = \frac{30}{25} Ra = 1.20 Ra$$

$$\text{Waste, } W = \frac{100(P_1 - P_2)}{(100 - P_2)} = \frac{100(75-70)}{(100-70)} = 16.7\%$$

(a) Calculations based on adjusting percent passing No. 4

Sieve	3/4"	1/2"	3/8"	#4	8	30	100	200
Spec.	100	80-100	70-90	55-73	40-55	20-30	10-18	4-10
% Pass, Pa	100	95	85	70	53	31	16	9
% Ret, Ra	0	5	15	30				
Adj % Ret, Rb	0	6	18	37				
Adj % Pass, Rb	100	94	82	63	48	28	14	8.1

$$Pb = \frac{P_2}{P_1} Pa = \frac{28}{31} Pa = 0.90 Pa \qquad Rb = \frac{R_2}{R_1} Ra = \frac{37}{30} Ra = 1.23 Ra$$

$$\text{Waste, } W = \frac{100(P_1 - P_2)}{100 - P_2} = \frac{100(70-63)}{100-63} = 18.9\%$$

(b) Calculations based on adjusting percent passing No. 30

Figure A-8– Adjusting gradation by wasting

A.11 COMPUTING LABORATORY BATCH WEIGHTS–In the analysis of aggregates for a given mix design, the final operation is the computation of the laboratory batch weights. It is convenient to use the same weight of aggregate in each batch of a trial mix series; in this way the only variable is the amount of asphalt to be added.

Thus, the first step is to determine the blend proportions of the sized fractions and mineral filler required to approximate the original trial blend. Any one of the methods illustrated previously may be used to determine the blend proportions that will yield the desired gradation and need not be repeated here. Figure A-9, however, shows the washed sieve analysis of the aggregate fractions and the blend proportions required for this example.

After the desired blend proportions have been determined, the computation of batch weights is simply a matter of calculating the amount of each aggregate fraction and asphalt cement required to yield a total amount of mix suitable for the preparation of individual test specimens. Figure A-10 shows the laboratory batch weights for this example. Note that all of the mixes use the same amount of aggregate and that the asphalt content is therefore expressed as a percent of the dry aggregate.

THE ASPHALT INSTITUTE HOT-MIX ASPHALT PAVING DESIGN AGGREGATE GRADATION ANALYSIS

Project: FI-005(4)
Location: Rex—East
Date: April 17, 1973

GRADATION OF AGGREGATE FRACTIONS

Aggregate Fractions	Percent Used	SIEVE SIZE—PERCENT PASSING								
		1″	¾″	½″	⅜″	No. 4	No. 8	No. 30	No. 100	No. 200
¾-⅜	100		100	65.4	1.8	0.4	—	—	—	—
⅜-4	100		100	100	100	1.9	0.9	0.3	—	—
4-8	100		100	100	100	100	1.8	0.8	0.6	0.2
-8	100		100	100	100	100	100	48.6	22.1	10.0
MF	100		100	100	100	100	100	100	95	74

COMBINED GRADATION FOR BLEND—TRIAL NO. 1

Aggregate Fractions	Percent Used	SIEVE SIZE—PERCENT PASSING								
		1″	¾″	½″	⅜″	No. 4	No. 8	No. 30	No. 100	No. 200
¾-⅜	20		20	13.1	0.4	—	—	—	—	—
⅜-4	20		20	20	20	0.4	—	—	—	—
4-8	18		18	18	18	18	0.3	0.1	0.1	—
-8	38		38	38	38	38	38	18.5	8.4	3.8
MF	4		4	4	4	4	4	4	3.8	3.0
Blend Grading			100	93.1	80.4	60.4	42.3	22.6	12.3	6.8
Desired Grading			100	90.0	80.0	60.0	42.0	23.0	12.0	7.0

COMBINED GRADATION FOR BLEND—TRIAL NO. —

Aggregate Fractions	Percent Used	SIEVE SIZE—PERCENT PASSING								
		1″	¾″	½″	⅜″	No. 4	No. 8	No. 30	No. 100	No. 200
Blend Grading										
Desired Grading										

Jones

Inspector

Note: To convert from inches to millimeters multiply by 25.4.

Figure A-9—Work sheet for analysis of aggregate blends

THE ASPHALT INSTITUTE HOT-MIX ASPHALT PAVING DESIGN LABORATORY BATCH WEIGHTS

Project: FI-005(4)
Location: Rex–East
Date: April 17, 1973

Lab. Nos.: 73-1071; 73-1072; 73-1073;

Trial Mix Series: 3-A Cold Aggregate Blend: CA-57% FA-39% Filler 4%

COMPUTED GRADATION OF HOT AGGREGATE BLEND

Sieve	1″	¾″	½″	⅜″	No. 4	No. 8	No. 30	No. 100	No. 200
% Pass		100	93.1	80.4	60.4	42.3	22.6	12.3	6.8

AGGREGATES

Fraction Used	Percent Used	Individual Weights (grams)	Accumulative Weights (grams)
¾-⅜	20	240	240
⅜-4	20	240	480
4-8	18	216	696
-8	38	456	1152
MF	4	48	1200
Total	100	1200	

ASPHALT CEMENT

Mix No.	Percent Used	Individual Weights (grams)	Accumulative Weights (grams)
1	4.0	48	1248
2	4.5	54	1254
3	5.0	60	1260
4	5.5	66	1266
5	6.0	72	1272

Jones

Inspector

Note: To convert from inches to millimeters multiply by 25.4.

Figure A-10–Work sheet for computing laboratory batch weights

INDEX

D

E

F

G

H

I

J

K

L

M

O

P

R

S

T

U

V

W

THE ASPHALT INSTITUTE

EXECUTIVE OFFICES AND RESEARCH CENTER
Asphalt Institute Building
College Park, Maryland 20740

The Asphalt Institute is an international, nonprofit association sponsored by members of the petroleum asphalt industry to serve both users and producers of asphaltic materials through programs of engineering service, research and education. Membership is limited to refiners of asphalt from crude petroleum, and to processors manufacturing finished paving asphalts and/or non-paving asphalts but not starting from crude petroleum. Institute Members provide quality products and advocate quality construction and timely maintenance. A total of 57 Members have headquarters in offices in:

United States	38	Europe	7
Canada	6	Middle East	4
South America	1	Asia	1

MEMBERS OF THE ASPHALT INSTITUTE

(As of August 1, 1979)

ALLIED MATERIALS CORPORATION, Oklahoma City
AMOCO OIL COMPANY, Chicago
ARCO PETROLEUM PRODUCTS COMPANY., Philadelphia
ASFALTOS ESPANOLES, S.A., Madrid, Spain
ASHLAND PETROLEUM CO., Ashland, Kentucky
ASPHALT MATERIALS, INC., Indianapolis
BP OIL LIMITED, Montreal, Quebec, Canada
BRITISH PETROLEUM COMPANY, LIMITED, THE, London, England
BYERLYTE COMPANY OF KOPPERS COMPANY, INC., Pittsburgh
CENEX, Laurel, Montana
CHEVRON U.S.A. INC., San Francisco
COMPANIA ESPANOLA DE PETROLEOS, S.A., Madrid, Spain
CONOCO INC., Houston
DELTA REFINING COMPANY, Memphis, Tennessee
DOUGLAS OIL COMPANY OF CALIFORNIA, Los Angeles, Costa Mesa
EDGINGTON OIL COMPANY, INC., Long Beach, California
EGYPTIAN GENERAL PETROLEUM CORP., THE, Cairo, Egypt
ELF FRANCE, Paris, France
ENERGY COOPERATIVE, INC., Chicago
EXXON COMPANY, U.S.A., Houston
EXXON CORPORATION, New York
GULF OIL CANADA LIMITED, Toronto, Ontario, Canada
HUNT OIL COMPANY, Dallas
HUSKY OIL MARKETING LTD., Calgary, Alberta, Canada
IMPERIAL OIL LIMITED, Toronto, Ontario, Canada
INDUSTRIA ITALIANA PETROLI SpA, Genoa, Italy
JORDAN PETROLEUM REFINERY COMPANY LTD., Amman, Jordan
KERR-McGEE REFINING CORPORATION, Oklahoma City
KOCH ASPHALT COMPANY, Wichita, Kansas
LAGOVEN, S.A., Caracas, Venezuela
LAKETON ASPHALT REFINING, INC., Evansville, Indiana
MACMILLAN RING-FREE OIL CO., INC., El Dorado, Arkansas
MARATHON OIL COMPANY, Findlay, Ohio
MOBIL OIL CORPORATION, New York
MOBIL OIL CORPORATION—International Division, New York
NATIONAL IRANIAN OIL COMPANY, Tehran, Iran
NATIONAL REFINERY LIMITED, Karachi, Pakistan
NEWHALL REFINING CO., INC., Newhall, California
AB NYNAS PETROLEUM, Stockholm, Sweden
OXNARD REFINERY, Oxnard, California
PAZ OIL COMPANY LTD., Haifa, Israel
PETROFINA CANADA LTD., Montreal, Quebec, Canada
PHILLIPS PETROLEUM COMPANY, Bartlesville, Oklahoma
POWERINE OIL COMPANY, Santa Fe Springs, California
SEMINOLE ASPHALT REFINING, INC., St. Marks, Florida
SHELL CANADA LIMITED, Toronto, Ontario, Canada
SHELL INTERNATIONAL PETROLEUM COMPANY LIMITED, London, England
SHELL OIL COMPANY, Houston
SOUTHLAND OIL COMPANY, Jackson, Mississippi
STANDARD OIL COMPANY, THE (Ohio), Cleveland
SUN PETROLEUM PRODUCTS COMPANY, Philadelphia
TOSCO CORPORATION, El Dorado, Arkansas
TOTAL PETROLEUM, INC., Alma, Michigan
U.S. OIL & REFINING COMPANY, Tacoma, Washington
VULCAN REFINING COMPANY, Cordova, Alabama
WARRIOR ASPHALT COMPANY OF ALABAMA, INC., Tuscaloosa, Alabama
WITCO CHEMICAL CORPORATION—Golden Bear Division, Los Angeles; **Pioneer Division**, New York

ASPHALT INSTITUTE ENGINEERING OFFICES

(As of August 1, 1979)

EASTERN DIVISION

WASHINGTON, D.C. 20014—4733 Bethesda Ave., (301) 654-5324, Connecticut, Delaware, District of Columbia, Kentucky, Maine, Maryland, Massachusetts, New Hampshire, New Jersey, New York, Ohio, Pennsylvania, Rhode Island, Vermont, Virginia, West Virginia

NEW YORK, N.Y.—(Address: 20 Evergreen Place, East Orange, N.J. 07018), (201) 676-3838, New York City, Westchester Co., Long Island, Delaware, New Jersey

SYRACUSE, N.Y. 13066—7000 E. Genesee St., Fayetteville, N.Y. (315) 445-1333, New York State except New York City, Westchester County, Long Island

BOSTON, MASS.—(Address: 599 North Ave., Door 9, Wakefield 01880), (617) 245-6046, Connecticut, Maine, Massachusetts, New Hampshire, Rhode Island, Vermont.

HARRISBURG, PA. 17109—13 North Progress Ave., (717) 545-4713, Pennsylvania

COLUMBUS, OHIO 43215—41 S. High St., Suite 101, (614) 224-3265, Ohio

LOUISVILLE, KY. 40207—4050 Westport Rd., (502) 895-6966, Kentucky, West Virginia

RICHMOND, VA. 23230—2120 Staples Mill Road, (804) 353-3826, District of Columbia, Maryland, Virginia

SOUTHERN DIVISION

JACKSON, MISS. 39206—415 Comet Drive, Suite 3, (601) 981-3417, Alabama, Arkansas, Florida, Georgia, Louisiana, Mississippi, North Carolina, Oklahoma, South Carolina, Tennessee, Texas (except El Paso area)

AUSTIN, TEXAS 78723—7701 Cameron Road, Suite 101, (512) 454-7927, Eastern Texas

NORTH LITTLE ROCK, ARK. 72116—P.O. Box 4007 (5318 John F. Kennedy Blvd.), (501) 758-0484, Arkansas, Oklahoma

MONTGOMERY, ALA. 36104—800 South McDonough St., (205) 264-3588, Alabama, Tennessee (except Memphis)

TALLAHASSEE, FLA. 32303—2639 North Monroe, Sta. 15, (904) 385-7822, Florida, Georgia

RALEIGH, N.C. 27605—2016 Cameron St., (919) 828-5998, North Carolina, South Carolina

NORTHERN DIVISION

CHICAGO, ILL. 60018—2400 E. Devon Ave., Suite 254, Des Plaines (312) 297-5907, Colorado, Idaho, Illinois, Indiana, Iowa, Kansas, Michigan, Minnesota, Missouri, Montana, Nebraska, North Dakota, South Dakota, Utah, Wisconsin, Wyoming

INDIANAPOLIS, IND. 46240—1010 E. 86th St., (317) 846-9146, Indiana

MINNEAPOLIS, MINN. 55422—6100 Golden Valley Rd., Suite 113, (612) 546-2311, Minnesota, Wisconsin

HELENA, MONTANA 59601—1111 North Rodney, (406) 442-9100, Idaho, Montana, North Dakota

JEFFERSON CITY, MO. 65101—400 East High St., Suite 305, (314) 635-1181, Missouri (except St. Louis County), Kansas

OMAHA, NEBR. 68132—6901 Dodge St., Room 105, (402) 556-4811, Nebraska, Iowa, South Dakota

DENVER, COLO. 80215—1401 Saulsbury, Suite 205, (303) 238-3411, Colorado, Utah, Wyoming

LANSING, MICH. 48912—416 Frandor Ave., Suite 205, (517) 351-6543, Michigan

SPRINGFIELD, ILL. 62704—1035 Outer Park Dr., Suite 413, (217) 787-7058, Illinois (outside of Chicago), St. Louis County, Missouri

PACIFIC COAST DIVISION

SACRAMENTO, CALIF. 95815—1796 Tribute Road, (916) 922-6507, Alaska, Arizona, California, Hawaii, Nevada, New Mexico, Oregon, Western Texas (El Paso area), Washington

LOS ANGELES, CALIF.—(Address: 4201 Long Beach Blvd., Suite 230, Long Beach 90807), (213) 427-7985, Southern California

PHOENIX, ARIZ. 85016—3625 N. 16th St., (602) 266-2516, Arizona, New Mexico, Western Texas

PORTLAND, ORE. 97005—Park Plaza West, Suite 322, 10700 SW Beaverton-Hillsdale Highway, Beaverton, (503) 643-4111, Oregon

OLYMPIA, WASH. 98501—120 Union Avenue Building, (206) 943-0320, Alaska, Washington

THE ASPHALT INSTITUTE

Divisional Areas and Offices in the United States

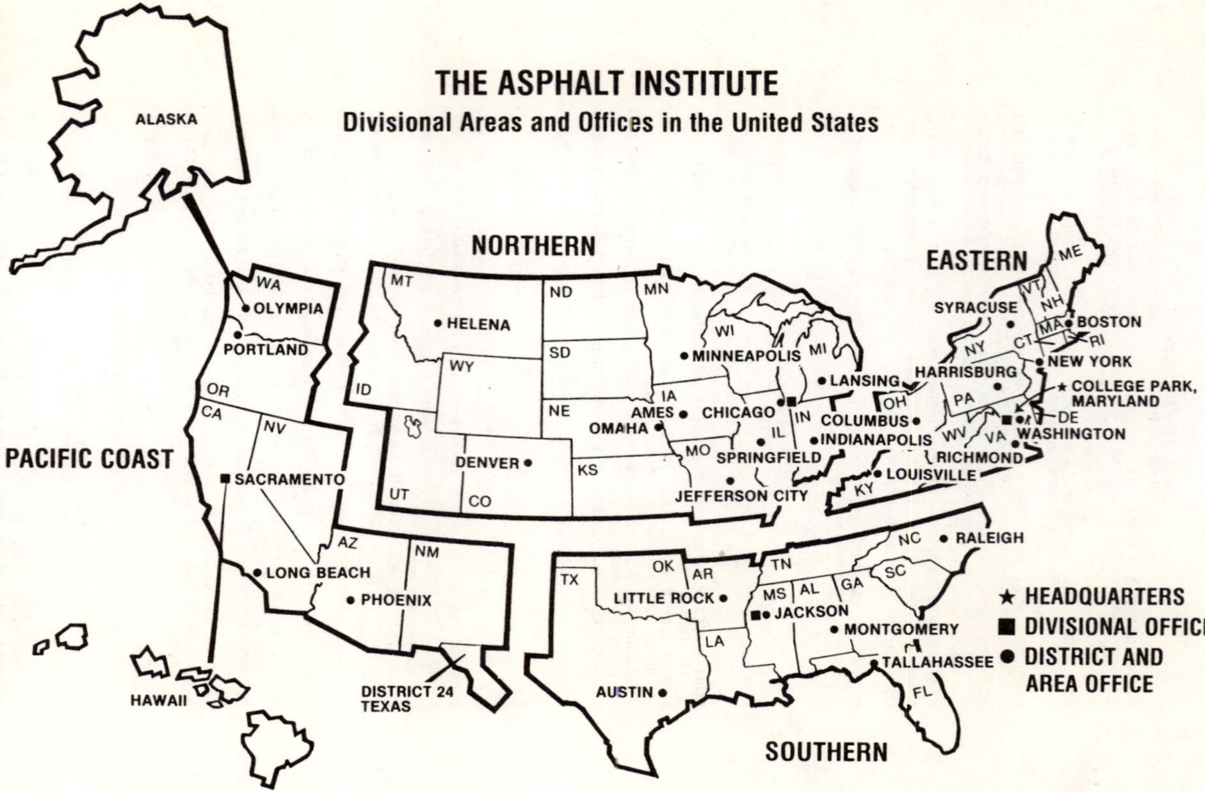

As of July 1, 1979

4thEdn374
2ptg778
3ptg379
L5m279
L10m979

Printed in U.S.A.